Frederic Schiller Cozzens

Yachts and Yachting

Frederic Schiller Cozzens

Yachts and Yachting

ISBN/EAN: 9783337405205

Printed in Europe, USA, Canada, Australia, Japan

Cover: Foto ©berggeist007 / pixelio.de

More available books at **www.hansebooks.com**

WITH OVER ONE HUNDRED AND TEN ILLUSTRATIONS

BY

FRED. S. COZZENS
AND OTHERS

CASSELL & COMPANY, Limited,
739 & 741 BROADWAY, NEW YORK.

CONTENTS.

THE HISTORY OF AMERICAN YACHTING. BY CAPTAIN R. F. COFFIN.
    I. EARLY DAYS OF THE NEW YORK YACHT CLUB,   11
    II. FROM 1859 TO 1870,   27
    III. THE INTERNATIONAL PERIOD,   43
    IV. FROM 1871 TO 1876,   58
    V. FROM 1876 TO 1878,   73
    VI. FROM 1878 TO 1885,   117
THE MAYFLOWER AND GALATEA RACES OF 1886. BY CHARLES E. CLAY,   103
AMERICAN STEAM YACHTING. BY EDWARD S. JAFFRAY,   115
BRITISH YACHTING. BY C. J. C. MCALSTER,   141

LIST OF ILLUSTRATIONS.

Mayflower and Galatea Contest,   Frontispiece.

	PAGE.		PAGE.
Adrienne,	66	Fleetwing,	27
Aida,	131	Fortune,	63
Alpha,	23	Frolic,	53
America,	22	Galatea,	106, 141
America's Cup, The,	108	Gertrude,	158
Atalanta,	119	Gimcrack,	15
Athlon,	74	Gitana,	82
Atlantic,	93	Gracie,	39
Bedouin,	87	Grayling,	62
Bianca,	44	Halcyon,	32
Buttercup,	153	Hassan Steam Launch,	117
Camilla,	117	Henrietta,	20
Carlotta,	147	Hope,	56
Clytie,	61	Hornet,	17
Constance,	149	Intrepid,	55
Comet,	34	Ione,	151
Corsair,	116	Irex,	143
Course, The (Mayflower and Galatea Races),	104	Isis,	59
		Julia,	19
Crocodile,	78	La Coquille,	14
Cygnet,	11	Lorna,	158
Dawn,	148	Lucky,	25
Diagram of Yachts,	105	Madeleine,	28
Diane,	152	Madge,	75
Dreadnaught,	33	Maggie,	88
Egeria,	156	Magic,	48
Electra,	122	Maria,	18, 38
Enchantress,	40	Marjorie,	150
Estelle,	60	Maud,	38
Fanita,	67	Mayflower,	107
Fanny (sloop),	47	Miranda,	144
Fanny (Boston),	51	Mischief,	69

LIST OF ILLUSTRATIONS.

	PAGE.		PAGE.
Mist,	16	Sappho,	31
Montauk,	70	Sentinel, .	121
Namouna,	135	Shadow, .	76
" Deck of the,	134	Spray,	13
Nooya, . .	126	Stiletto,	133
" Deck View, .	127	Stranger, .	91
Nourmahal, . .	135	Sunbeam,	129
" The Bridge, .	137	Sybil, .	12
" Working Drawings,	136	Tara,	154
" The Officers' Room.	132	Thetis,	54
Orienta,	131	Uledia,	146
Oriva, .	92	Una,	21
Painting the Boat,	130	Utowana,	118
Palmer,	52	Vesta,	36
Pastime, .	120	Viking,	125
Polynia,	133	Vixen,	35
Priscilla, .	95	Wanda, .	126
Promise, .	128	Wanderer,	37
Queen Mab.	145	Water Witch, .	157
Radha,	133	Whisper, .	135
Ray,	24	White Wing, ...	45
Rebecca, .	30	Yacht Stations of the British Isles,	142
Samoena, .	155	Zoe,	80

THE HISTORY OF AMERICAN YACHTING.

THE HISTORY OF AMERICAN YACHTING.

BY CAPTAIN R. F. COFFIN,

Author of "The America's Cup," "Old Sailor Yarns," etc., etc.

I.

EARLY DAYS OF THE NEW YORK YACHT CLUB.

The history of the New York Yacht Club, for the first sixteen years of its existence at least, is practically the history of American yachting. For the first five years it was the only yacht club in the United States. The Southern yacht club, with its head-quarters at New Orleans and its racing course on Lake Ponchartrain, was the second club organized, but it was purely a local organization; its yachts were small open boats, and it had no real national influence. The third club was the Neptune, at the Highlands, organized in 1850, one year later than the Southern; but it was rather a fishing than a yachting club, and like the Southern, its boats were for the most part open centerboards of small size. It may have had an occasional race on the Shrewsbury River, but it was simply a summer organization, and, except in name, had none of the characteristics of a regular yachting organization. The Carolina club, with head-quarters at Wilmington (if I am not mistaken), North Carolina, ranks fourth in point of age, but was of no importance except, perhaps, locally. It was organized in 1854, but it was not until the Brooklyn Yacht Club was organized in 1857, that there was anything like a yacht club, in the present significance of the term, in all the United States. It had its head-quarters at the head of Gowanus Bay, and was naturally an association of the gentlemen who had, for several years previous, made this the anchorage for their pleasure craft. This locality was then, and for a long time afterwards continued to be, the principal place for boat sailing in the city of Brooklyn. The Brooklyn was from the first a real yacht

MR. W. EDGAR'S "CYGNET."

MR. C. D. MILLER'S "SYBIL."

club, as was also the Jersey City, which was organized in 1858, and which was the natural associating together of gentlemen who were in the daily habit, during the summer, of sailing their boats from along the shores of Communipaw Bay. From this time on, for seven years, New York was the home of yachting, no club being organized in any other waters; but in 1865, the Boston club was organized, and this is the oldest club in the New England States. By this time, the Brooklyn club had grown large enough to put out an off-shoot, and a few of its members withdrew from it and organized the Atlantic, which is the eighth yachting organization in this country, having been organized in 1866. In 1867, the Columbia club was organized on the West side of the town, having its head-quarters at the foot of Christopher street—if I remember aright—and, like the Brooklyn and Jersey City, it was a natural association of gentlemen who had been in the habit of keeping their boats there, and sailing thence on the Hudson River. Its boats were open centerboards for the most part, but it has always been a respectable and, with some few exceptional years such as occur in the history of most of the yachting organizations, it has been prosperous. It has a nice club house and excellent anchorage now, at the foot of West Eighty-fourth street. It was organized in 1867, and in the same year the San Francisco club, at the Golden Gate, came into existence. The South Boston was started in 1868, the second in New England, and the Bunker Hill and Portland in 1869—a total of but fifteen yacht clubs in all the United States. In that year Mr. Ashbury, with the British schooner *Cambria*, came to race for the America's Cup, and this gave an impetus to yachting, which, aided by other causes, has continued to the present time, and has caused the multiplication of clubs, so that there are now over one hundred and twenty of these organizations, and they are increasing rapidly in all parts of the country, each of the great lake ports having its club; while in the New England States they are more numerous than in any other section of the country, and Boston, to-day, is more of a yachting center than New York itself.

Still, the history of the New York Yacht Club is to a great extent the history of American yachting; for down to the year 1885 no other club ever attempted anything more than mere local effort. Each organization had its club house and anchorage, its regattas, or more properly speaking, matches, over its regular course, one or more times a year, and that was all.

Very few had yachts large enough, in number sufficient, to essay a squadron cruise previous to 1870. I doubt if any of them except the Brooklyn had, and as for ocean races, or private matches, for valuable prizes, all that sort of thing was left alone to the New York club, which from the first has displayed an enterprise and a boldness worthy of the great city of its home and name.

I don't think that the early history of the club events has ever been written. I, certainly, have never seen or heard of anything of the kind, and then I think a brief sketch will be of interest. In no other

MR. H. WILKES' "SPRAY."

manner can the early history of American yachting be as well told.

Organized July 30, 1844, the New York Yacht club had its first regatta July 16, 1845. Its first rating for allowance of time was per ton, of custom house measurement, and the allowance was forty-five seconds per ton. Its first course was from a stake-boat off Robbins Reef, to a stake-boat off Bay Ridge, L. I.; thence to a stake-boat off Stapleton, S. I.; thence to and around the Southwest Spit buoy, returning over the same course. Presumably, every yacht in the club started, for it was the first regatta ever sailed in this country. I don't know whether or not this was the case, and it is of no importance; but they had a very respectable fleet in point of number, although none of the starters were of any great size, the largest being but forty-five tons; about sixty feet long perhaps.

Evidently the schooner was the favorite rig, as it has been ever since, with some exceptions, and as it will probably be again, unless the steam yacht take the schooner's place and the sailing yacht, kept purely for racing, be confined to the single-stick boats. The schooner rig, however, is so much handier, that it is sure to be preferred for a vessel kept solely for pleasure sailing.

Let us see who the starters were in this first regatta, and who owned them: Schooner *Cygnet* (45), Mr. W. Edgar; *Sybil* (42), Mr. C. B. Miller; *Spray* (37), Mr. H. Wilkes; *La Coquille* (27), Mr. John C. Jay; *Minna* (30), Mr. J. Waterbury; *Gimcrack*, Mr. J. C. Stevens.

Sloops: *Newburgh* (33), Mr. H. Robin-

MR. JOHN C. JAY'S "LA COQUILLE."

son; *Adda* (17), Mr. J. Rogers; *Lancet* (20), Mr. G. B. Rollins.

These, then, were the men, and their yachts with which the New York Yacht Club went into business in 1845. The only yachts timed at the finish were the *Cygnet*, which won, in 5h. 23m. 15s., the *Sybil*, coming second, in 5h. 25m. 25s.; and the *Gimcrack*, in 5h. 30m. 30s. The prize was as good a cup as could be purchased with the entrance money, which was, I think, $25 for each yacht. Schooners and sloops were classed together, and there was no allowance for difference of rig. The schooner had no foretopmast, and of course that stick-breaking sail, the jib topsail, was unknown, as was also the club topsail. These were later inventions. In fact I doubt whether the schooners sported such a thing as a staysail on this occasion. The sloop had a short bowsprit and short topmast and no jib-boom.

The regatta was a great event, and was witnessed by thousands of people, all New York, who could get there, being on the water. Now-a-days an ordinary club regatta attracts few besides club members, and old yachtsmen shake their heads gloomily, and lament the decadence of American yachting, saying that all interest in the sport is dying out; but in point of fact, there was never as much interest as at present; only now it is diffused, then it was concentrated. In those early years of American yachting, the regatta day, or days, of the New York Yacht Club were almost general holidays among the men of large business, brokers and jobbers; and every craft that could float, from the skiff to the large excursion steamer, was brought into requisition for spectators.

Before the next summer arrived, the club had built itself a house at the Elysian Fields, Hoboken; and for more than twenty years the start and finish of its races were off this place, thousands of

MR. J. C. STEVENS' "GIMCRACK."

people congregating there to see the finishes. The house then built still stands, and is now used as a club head-quarters by the New Jersey Yacht Club, which sails its races over very nearly the same course as that adopted in 1846 by the New York club, on the occasion of its second match. largest schooner was Mr. J. H. Perkins' *Coquette* (76), and the largest sloop was Mr. L. Depau's *Mist* (44). The only yacht which made the course inside of the limit of eight hours was the sloop *Mist*, which did it in 7h. 37m., winning the prize, this time offered by the club, and which was of the value of $200; and this is called in the club annals its "first annual regatta." Why, I don't know, since, as we have seen, there had been one during the previous summer. The allowance was the same as before — forty-five seconds to the ton — and schooner and sloop, all went in together, the sloop as we have seen, getting the best of it. The club had another race the next day, July 18, 1846; but the two days are properly classed by the club as one regatta. This time the starters were: Schooners, *Gimcrack*, *Hornet*, *Minna*, *Brenda*, *Cygnet*, *Siren* and *Coquette*, and the sloops, *Pearsall*, *Mist*, *Ann Maria* and *Dart*. I think that the *Pearsall*, *Ann Maria* and *Dart* were working vessels, allowed to come in on even terms with the club boats. The course was the same, and it was for many years the regular club course. The *Gimcrack*, finished first, *Mist* second, *Hornet* third, *Dart* fourth; Mr. A. Barker's schooner *Hornet* (25), winning, on allowance of time, a piece of silver valued at $200.

"MIST."

This course was from a stake-boat off the Elysian Fields, to a stake-boat off Stapleton S. I.; thence to a stake-boat off the Long Island shore, and thence to the Southwest Spit, returning over the same course. For this, there started: the schooners *Lancet*, *Gimcrack*, *La Coquille*, *Minna*, *Brenda*, *Spray*, *Sybil*, *Cygnet*, *Pet*, *Northern Light*, *Siren* and *Coquette*, with the sloops *Newburgh* and *Mist*. The

There has been some little boasting on the part of the Seawanhaka club over a claim that they were the first to introduce "Corinthian" racing, and so fearful have

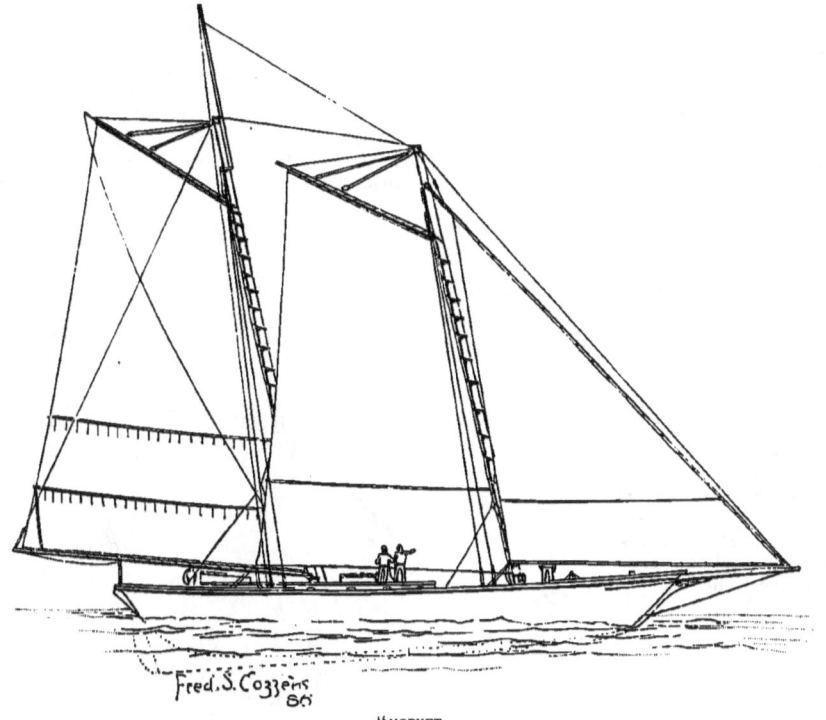

"HORNET."

its members been that their superiority in this respect would be lost sight of in the now almost general adoption of this method, either in whole or in part, that some years ago they tacked the word "Corinthian" on to their originally beautiful Indian name "Seawanhaka," making a clumsy and cumbersome title out of their first extremely appropriate name; and I suppose few, if any, of these young gentlemen are aware that on the 6th day of October, 1846, the New York club sailed a match for a cup, subscribed for by members, the rule being—I quote literally—"none but club members allowed to sail or handle the boats, but each yacht may carry a pilot." They could not even have their sailing-master, except as pilot, which is further in the "Corinthian" line than even the Seawanhakas have ever gone.

The course was from a stake-boat off the Elysian Fields to a stake-boat off Fort Washington Point; thence to a stake-boat anchored in the Narrows, returning to the place of starting, a distance of forty miles, with an allowance of 25 seconds per ton.

It was in this race that the sloop *Maria*—afterwards celebrated—made her first appearance. She was 160 tons, and was owned and sailed by Mr. John C. Stevens, who was then commodore of the club. The other "Corinthians" were the sloop *Lancet*, and the schooners *Siren*, *Cygnet*, *Spray* and *La Coquille*. The wind was a strong breeze from southwest, and the *Maria* won, beating the *Siren* 58m. 15s. actual time.

Four days later, namely, October 10, 1846, the first ocean race ever sailed by yachts, came off. It was a match for $1,000 a side—pretty good that for a club only two years old—between the sloop *Maria* (and this time she is entered at 154 tons) and the schooner *Coquette*, the course being 25 miles to windward and return, the wind strong from the northeast, and the boats went from the buoy at the entrance of Gedney's Channel to a stake-boat off the south ends of the Woodlands. The *Maria* started with double-reefed mainsail and bonnet off of the jib, the schooner carrying all sail all through the race. Evidently the wind was too strong for the sloop; but it was a close race, the schooner winning in 6h. 35m. 30s.; the sloop 7h. 1m.

These were yachting years evidently, and next year, viz., 1847, they got at it early, the schooners *Sybil* (42), Mr. C. Miller, and the *Cygnet* (45), Mr. D. L. Suydam, sailing a match on the 25th of May, for $500 a side, over the regular club course, and the *Sybil* won.

On the 31st of May, 1867, which now-a-days is our great opening day, there was another match race sailed for $500 a side, between Mr. William Edgar's schooner *Cornelia* (90), and Mr. D. L. Suydam's schooner *Cygnet* (45), over the regular club course. The *Cornelia* grounded off Ellis' Island, going down, and, of course, the *Cygnet* won.

At the regular regatta, this year, which took place June 2, the sloop *Una*, afterwards so celebrated, and after which, it was thought by some, the sloop *Puritan* was modeled, sailed her first race. She was owned by Mr. James M. Waterbury, and was 39 tons.

For the first time two classes were made, showing that the yacht owners were being gradually educated in the ethics of the sport. There were three entries in the first, and six in the second class. The rigs, however, were not separated, and in the first class the schooners *Cornelia* and *Siren* were pitted against the sloop *Maria*.

Then there was a class for outside vessels, in which there were four starters, two schooners and two sloops. There were three different allowances, viz.: for first-class club boats, 35 seconds; for second-class club boats, 45 seconds; and for the outside craft, 40 seconds per ton. The wind was fresh from southwest, and the *Maria* and *Una*, in their respective classes, won as they pleased, while the sloop *Dart* (59), an outside boat, took the prize there. Now-a-days, a proposition to race a sloop against a schooner inferior in size, and without allowance for rig, would be laughed to scorn.

October 12 of this year, 1847, there was another "Corinthian" race, and this time over the regular club course. It was for a subscription cup, the yacht *to be manned and*

"MARIA."

sailed *exclusively by* members, allowing each yacht a pilot, and there started the schooners *Gimcrack, Dream, Spray, Cygnet, Siren,* and *Cornelia,* with the *Una.* Of course, the *Una* won, *Siren* second, *Spray* third.

The regatta called third on the club record took place June 6, 1848, the yachts being divided into two classes, but no separation of rig. There was a cracking breeze from the west-north-west, and the *Maria* was dismasted, a bad habit which she contracted in her youth, and never recovered from. She was finally altered to schooner, because, among other reasons, it had been found impossible to hold her stick in her. She was a boat of enormous beam and great initial stability, and her sail spread was something prodigious. The accident on this occasion took place between Jersey City and Hoboken, when she was at the head of all the fleet except the *Cornelia,* and was gaining very rapidly on her. The *Maria* seems to have been constantly shrinking in size, for at this regatta she is entered at 118 tons, quite a drop from 160, at which figures she sailed her first race. The winners were, in the first-class schooners, *Cornelia* and *Siren,* and in the second-class, *Cygnet* and *La Coquille.* Thus it will be seen that in the process of evolution the club had come to two prizes in each class.

October 26, 1848, there was a kind of experimental match proposed, showing that there had been some dissatisfaction, and that there was a reaching out for something better. It was a "Corinthian" match, members to steer and man the yachts, and was for three pieces of plate subscribed for by members. One of these was to go to the best schooner, one to the second schooner, and one to the best sloop. This, then, was the first race in which the two rigs had been separated. The starters were the schooners *Sybil* (37), *Siren* (60), *Breeze* (74), *Cornelia* (75), over the regular club course. The allowances were: 35 seconds for over 40 tons; 45 seconds for 40 tons and under classes of schooners. The wind failed, and the race was not finished, and had to be resailed November 3, when

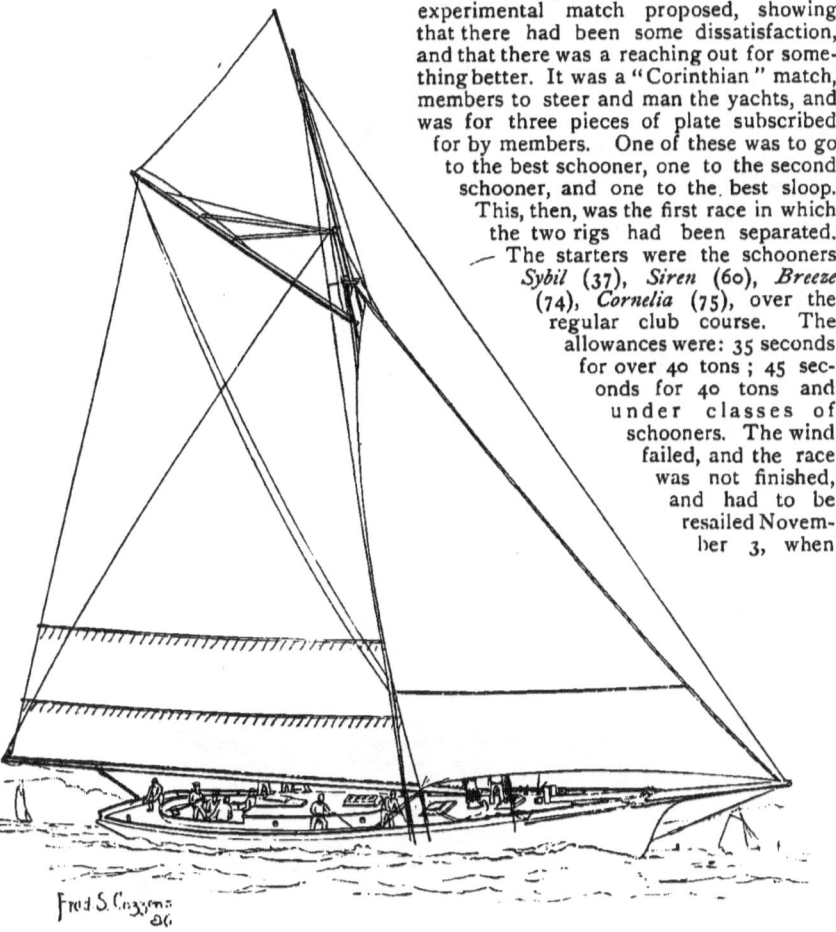

"JULIA."

only the schooner *Sybil* and *Cornelia* came to the line and went over the course. The *Sybil* won, and the *Cornelia* captured the second prize. Evidently, owners were getting better acquainted with their boats and those opposed to them, and they do not seem to have relished defeat any more than their successors do now-a-days. There were no starters among the sloops, probably for the reason that as against the *Maria* and *Una* no vessel stood a chance, and it was labor lost, and money thrown away, to fit them for a race. Previous to this race of November 3, however, *viz.*, October 31, 1848, there was a match between the sloops *Una* and *Ultra* over the club course. The *Ultra* was 65 tons, the *Una* but 39 ; and size told in those days as now, and the allowance of 17 minutes 15 seconds which the *Una* got was not enough for her. The *Ultra*, then owned by Mr. C. B. Miller, won by 15 minutes.

Next year, 1849, there was no race until the regular June regatta, which occupied two days, June 5 and 7. On the first day the course was the regular one, but the second day the start was off Robbins' Reef, and the yachts went around the Sandy Hook lightship for the first time. The *Maria* came in ahead on the first day, but was disqualified on account of fouling the *Ultra*. On the second day the prizes were $50 for each

"HENRIETTA."

class, no separation of rig; an allowance of 35 seconds a ton for all over 50 tons, and 45 seconds for all yachts 50 tons and under. There started in the first class the schooners *Cornelia* and *Siren*, and the sloop *Maria*, and in the second class only the schooner *Sybil*. Evidently a voyage around the lightship was looked upon with some distrust by the yachtsmen of 1849. The *Maria* came in ahead, but lost on allowance of time to the schooner *Cornelia*. The *Sybil*, of course, had a walk over.

In those days, however, this going around the lightship was considered a great feat, and by many, a most imprudent proceeding. Still the more adventurous did not think so, and October 13 the schooners *Cornelia* and *Breeze* sailed a match over this course, starting from Robbins Reef.

the club course, and the second day around the lightship, starting from Robbins Reef. On the second day the *Maria* sprung the head of her mast when near

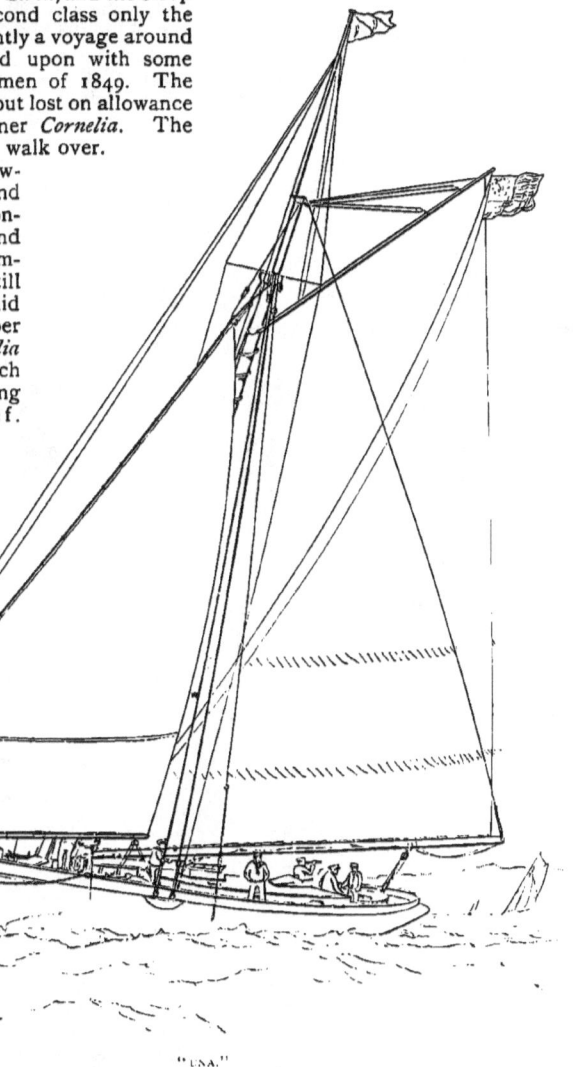

"UNA."

Sandy Hook, bound out, and had to give up. By this time the yachtsmen had got to protesting quite lively against each other, and regatta committees had plenty of work after the races, deciding questions of violation of rules. In 1851 there was

There was a fresh breeze from north by east to east by north, and the *Breeze* carried away her bowsprit, and the *Cornelia* sprung her mast, but came home a winner.

Next year there was again two days' racing, June 6 and 7, the first day over

also two days' racing over these same two courses, with six starters on the first day, and but four on the second. This, however, was a memorable year, for this was the summer that the schooner *America* came out and sailed across the ocean.

On Friday, May 9, 1851, at a general meeting of the Royal Yacht Squadron, a cup valued at $100 was offered for competition by yachts of all nations, the course being around the Isle of Wight, starting and finishing at Cowes; and for this race the *America* started against eight other schooners and nine cutters. The result is too well known to require more than a passing allusion. The *America* came in winner with loss of jib-boom; the three first arrivals being the *America* (170), at 8.37; *Aurora*, cutter (47), at 8.45; the *Bacchante*, cutter (80), 9.30. The reason of the *Aurora* getting so well placed at the finish was that the end of the match was a mere drift, but had there been an allowance for difference of size, the *Aurora* would have been beaten by less than two minutes, although, when passing the Needles, on the return, she was full eight miles astern, and the rest of the fleet out of sight astern.

I may say a word in passing of the other race sailed by the *America* in British waters, before she was sold by her American owners. This race is not so well known as the other for the cup which still bears her name. This was a match sailed August 28, 1851, with the schooner *Titania* (100). The course was from the Nab to a station twenty miles away, either to the leeward or windward, as the case might be. There the first part of the race ended, and £50 was to be awarded to the winner. The yachts were then to be again started, on the return, and another £50 depended on the finish. The wind at the start was fresh from northwest, increasing to a gale, and hauling to north by west. In the run off, the *America* beat the *Titania* four minutes twelve seconds, and in the beat back, she beat her 52m., thus winning the whole £100.

"AMERICA."

Nearly each year, as we have seen, brought something new in the history of American yachting, and 1852 was no exception to this rule, the yachts at the annual regattas, which took place June 3, being for the first time divided into three classes: over fifty tons; between fifty and twenty-five tons, and twenty-five tons and under. It was sailed over the regular club course, and there was no separation on account of rig. In the first class were the sloops *Una*, *Sybil*, and *Ultra*, and the schooner *Cornelia*. In the second class the sloop *Sport*, and in the third the sloop *Alpha* and schooner *Ariel*. The winners were the *Una*, *Sport*, and *Alpha*. Next day, June 4, the course was around the lightship from Robbins Reef, but the wind was so light that it had to be resailed, June 9, after another ineffectual trial, June 7, and the sloop *Silvie* (68), Mr. Louis A. Depau, won her first race. She afterwards became very famous as the first American sloop yacht which ever went across the ocean to England.

Next year, 1853, the programme for the annual regatta was the same as in the few previous years; a race over the old club course from the Elysian Fields for the first day, and presumably for the benefit of the public, and on the second day, a race around the lightship. Owners were chary of attempting the outside course, and there were but four starters, the sloops *Alpha* (17); *Sport* (26), and *Una* (54), with the schooner *Cornelia* (78). I think the *Una* won, but am not certain.

On August 9, of this year, 1853, there was another race at Cowes, Isle of Wight, in which an American yacht—the sloop *Silvie*—took part. The course was sixty-six and three-fourths miles long, and the starters were the cutters *Arrow* (102); *Julia* (111); *Aurora* (60); sloop *Silvie* (105); Swedish schooner *Aurora Borealis* (250); schooners *Alarm* (248); *Osprey* (59). The start was from an anchorage, and the *Silvie* was beaten by the cutter *Julia*, 6m. 38½s.

I think it was in the year 1854 that balloon sails, club topsails, etc., came into vogue, for a resolution of the club provided that "there should be no restrictions as to canvas that may be carried by yachts contending for prizes." It was evidently necessary to offer inducements for entries, as in the first class on the first day of the regatta, which was June 1, the starters were: two sloops and three schooners; in the second class the same, and in the third class four sloops. Over the outside course, two days later, the only yacht which came to the starting line was the sloop *Alpha* (17), and she started at 11.48 A.M., and finished at 7.12:30 P.M.

This year was memorable for the first

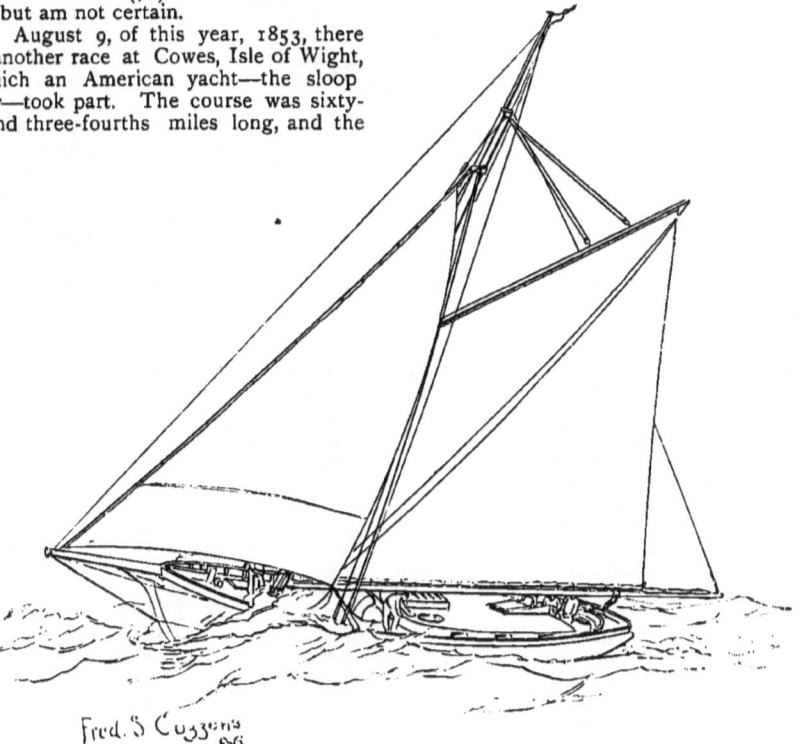

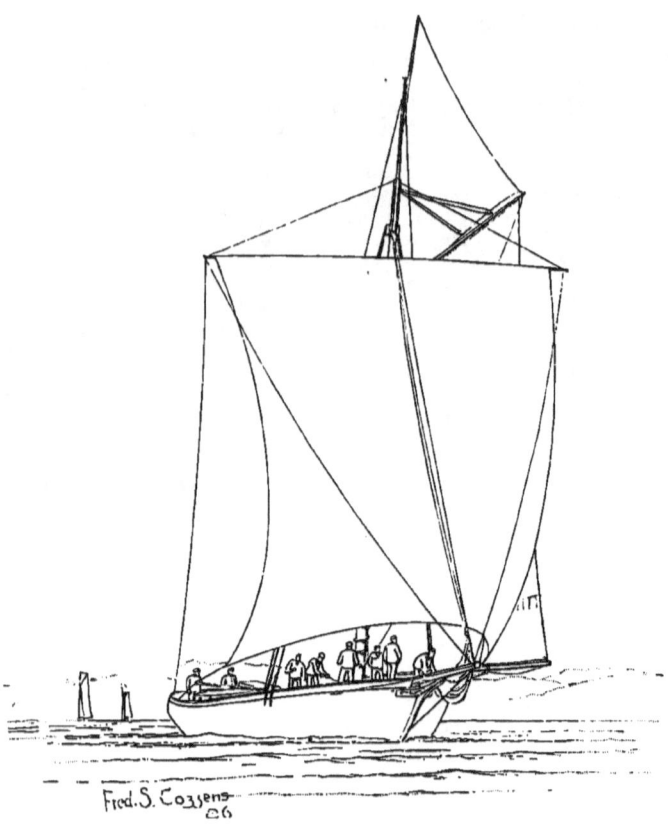
"RAY."

race the club ever sailed at Newport, and this was the first annual cruise of the club. The prize for the race at Newport, August 10, 1854, was $500 for yachts of any club, and $100 for working vessels; the course being what is still known as the club course at Newport, i.e., from off Fort Adams to and around the Block Island buoy and return, 45 miles. The starters were: the sloops *Maria* (116) [she seems to have received a different rating at each race]; *Julia* (80) [her first race]; *Una* (58), *Gertrude* (69), *Irene* (48), *America* (40), and *Ella Jane* [a working vessel] (89); schooners *Haze* (80), *Cornelia* (78), *Mystery* (46), *Spray* (37). The wind was northeast, and there was considerable sea. The sails were limited; sloops to mainsail and jib; schooners, to the three lower sails; and all started with the topmasts down. The *Maria* took the $500, and the *Ella Jane*, being the only working vessel, captured $100. She came in 49m. 28s. astern of the *Maria*, and had started 4m. 15s. after the *Maria*. The yacht, therefore, beat her 45m. 13s.

Next year, 1855, was the tamest annual regatta in the experience of the club. There was but one day's racing, a match over the old course, from the Elysian Fields; but on August 3, when the club was about to start on its second annual cruise, it had a regatta from Glen Cove Harbor, for a prize contributed by "citizens," ostensibly, but really by W. E. Burton, the popular actor, who at that time owned the place at Glen Cove now owned by Mr. S. L. M. Barlow, and who was a member of the club. The course was from off the steamboat dock to a stake-boat off Throggs' Point, thence to Matinnicock Point, and back to the place of departure, a distance of twenty-five miles, on an allowance of twenty-five seconds per ton. There were five schooners and fourteen sloops started, showing that in the natural process of evolution the sloop rig had come to be the favorite; largely influenced, I presume, by the splendid performances of the *Maria*, *Una*, *Julia*, and *Silvie*. Anyway, this was a land, or watermark in the history of the club; it was its first race on Long Island Sound, unless the previous year's race at Newport be considered as such. This year, also, at Newport, saw a variation of the previous year's programme, the yachts, on August 14, sailing a match from Fort Adams to Hop Island

and return, twenty-two miles, the first class, 60 tons and over; second, between 60 and 30 tons; and third, 30 tons and under. There were four starters in each class, two prizes of $200, and one of $100. Evidently this course did not suit as well as the other, for it has never been sailed over since by this club.

In 1856 the club took a new departure. There had been more or less dissatisfaction with the system of measurement; I may say there has been ever since, and probably always will be; and this year a change was made, and the allowance was based on sail area. Of course it was as unfair as possible, but was gotten up by those owners who did not care to carry balloon canvas, and wanted to penalize those who did. Like the present system of length and sail area, it was designed to benefit a particular class of yacht. The 1856 rule provided that yachts carrying less than 2,300 square feet of canvas should go in the third class, and their allowance of time over the club course should be 1½ seconds per square foot. The second class included yachts carrying 2,300 square feet and upwards, and the allowance was 1¼ seconds per foot. The first class included those yachts which had a sail area of 3,300 square feet and upwards, with an allowance of one second. The new regulation had one good effect; there was a larger entry at the annual regatta than for many years. The yachts were started by classes, as they stupidly are to this day in some of the clubs, and there was only one day's racing, the light-ship course being very unpopular.

The fact was, that the immense crowds which used to throng the Elysian Fields to witness the finishes, had much to do with this; the yachtsman of that day, resplendent in blue and gold, felt himself in the presence of this assembly of the populace a very much more important individual than he esteems himself now-a-days. All this was previous to the war, which made the wearing of a uniform very common, and which has caused this, the great club of the country, to discard it almost altogether, and relegate the blue and gold to the minor clubs.

This year was marked by another event, and that was that on August 8, 1856, the club sailed its first regatta at New Bedford over a course 32¼ miles long. Let me show how unequally this sail-

"LUCKY."

Fred. S. Coggens

area rule worked. Here are the entries for this race:

Name.	Tons.	Square feet.
Sloop *Silvie*,	100.0	4580.88
" *Widgeon*,	101.9	3502.44
" *Julia*,	83.0	3307.45
Schr. *Favorite*,	138.0	3983.20
" *Haze*,	87.2	3542.05
" *Twilight*,	73.6	3283.20

Thus, the *Widgeon*, a larger yacht than the *Silvie*, has to receive time from her, and she also receives time from the *Haze*, a vessel 14 tons smaller than herself, and with not as fast a rig, as a general rule. The *Julia* barely won, with the *Widgeon* second.

From this time on, until June 24, 1858, there was nothing out of the ordinary course of things in the history of the club, or in the history of American yachting. June 4, 1857, the annual regatta consisted of one race over the regular club course. August 13, 1857, there was another race at New Bedford, this seeming at this time, as always, a favorite place with the club. June 3, 1858, another race over the old club course for the annual event; but on June 24, 1858, there was a race around Long Island, and of course the first ever sailed. This was the race in which the late commodore, Mr. J. G. Bennett, then with Jr. after his name, became famous for taking a short cut through Plum Gut instead of through the Race, as provided by the articles of agreement. The start was from the Elysian Fields, down through the Narrows, and out by Sandy Hook; and the finish was at Fort Schuyler. The entrance fee was $50 for each yacht, and there was no restriction as to canvas. As this was rather a celebrated contest, I will give the entries in full:

	Name.	Owner.	Tons.
Schr.	*Haze*,	W. W. McVicker,	87.23
"	*Silvie*,	W. A. Stebbins,	105.04
"	*Favorite*,	A. C. Kingsland,	138.00
"	*Widgeon*,	Wm. Edgar,	101.09
Sloop	*Rebecca*,	J. G. Bennett, Jr.,	77.06
"	*Madgie*,	R. J. Loper,	99.05
"	*Una*,	W. B. Duncan,	67.05
"	*Minnie*,	S. W. Thomas,	59.14

At Fire Island the *Una* and *Rebecca* led: the *Rebecca* was first at Montauk Point; *Favorite* second, and two minutes behind. The *Minnie* protested against the *Rebecca* for going through Plum Gut, and she was ruled out. The *Silvie* won the schooner prize, and the *Minnie* took the sloop prize.

This year was wound up by a fall regatta on the 30th of October, the start for the first time being made from off Owls Head, as at present, and out around the lightship. It was the same as the present course, except that the finish was at the same place as the start, instead of as now, at buoy 15.

THE HISTORY OF AMERICAN YACHTING.

BY CAPTAIN R. F. COFFIN,

Author of "Old Sailor Yarns," "The America's Cup," etc., etc.

II.

FROM 1859 TO 1870.

In the first chapter, I brought the history of what may be called the public yachting —that in the races of the yachts of the New York club down to the close of the season of 1858, and it will be remembered that down to one year previous to this, no real yachting organization had been formed, and the Brooklyn club, which, properly speaking, was the second organization formed was of but little importance for the first seven years of its existence, not being an incorporated body until the year 1864. Its yachts were for the most part open centerboards, sloop or cat rigged, with perhaps a few cabin sloops of small size. In fact it was not until the election of Mr. Jacob Voorhis, Jr., as its commodore, which was, I think, in 1869, that it attained any prominence. That gentleman, then the owner of the schooner yacht *Madeleine* —a man of wealth, and a member of the New York club—brought with him many of the prominent yacht owners of that organization, and gave to the Brooklyn club, on its roll at least, a national importance. It is a matter of doubt, however, whether this added any real strength to it at all. The allegiance and sympathies of these men were with the parent club, and a few years later, they all withdrew from membership in the Brooklyn.

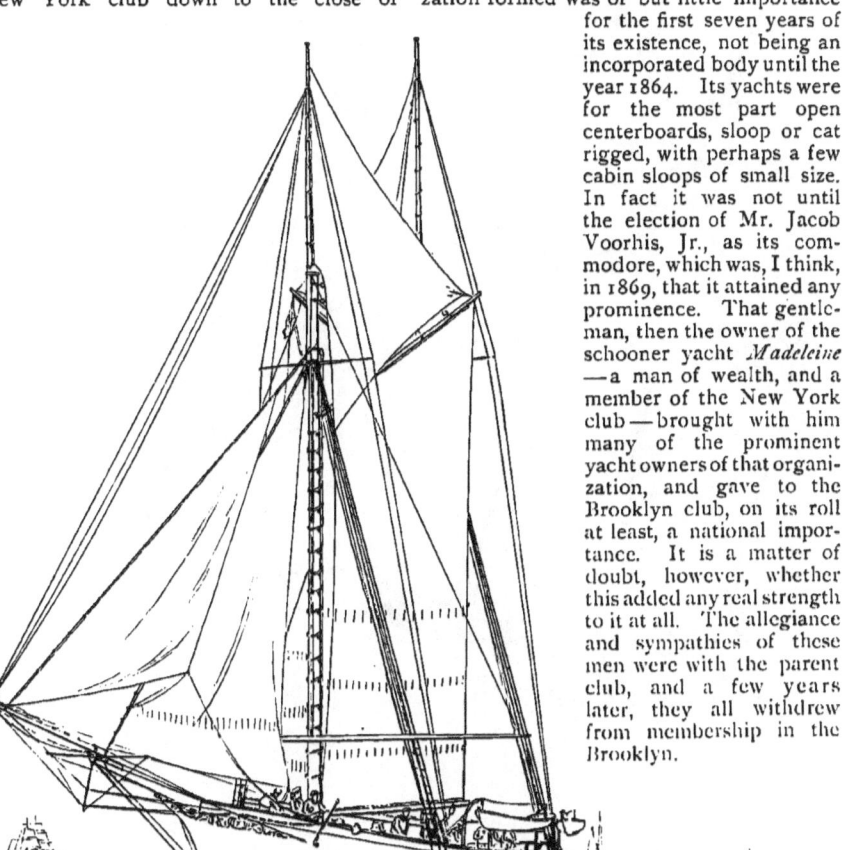

"FLEETWING."

So then, for some years at least, from the date of my last résumé, the history of the New York club was practically still the history of American yachting. Beside the public races at the regular regattas, and the private matches, there is a history of the sport, which, if the data were obtainable, would be found far more interesting than these, and that is the account of the private cruises and the afternoon sailing; these, after all, constituting the real enjoyment of the sport, to which the public races are merely incidental. It is these that make yachting the very prince of out-of-door sports. It is free from all the abuses and objections attaching to the turf, and must, from the very nature of things, always be the sport of gentlemen.

In the first place, none but the comparatively wealthy can own and run a vessel kept purely for pleasure sailing, and it is difficult to see how a man can expend his wealth in sport more profitably to himself, his friends, and the community. In the equipment and victualing of a yacht, all classes of the community receive a share, and the intimate friends of the owner receive that which is most valuable of all, the health-giving exercise and the fresh sea air which is its accompaniment, the owner himself getting in these ample return for all his outlay.

So, in all these years of which I have written, I can picture the splendid fleet, getting under-way each fine afternoon of the season, from off the Elysian Fields, and according as the tide served for a sure return in the early evening, sailing either down the Bay or up the Hudson River, the club, in those early days, being more fortunate than during its later years, when, on account of the

SCHOONER "MADELEINE." *

encroachments of the commerce of the port, its afternoon sails must always be made toward Sandy Hook. It was more fortunate also in another respect, that then it had a regular anchorage, with a club-house and landing near by.

After all, however, the cruises up the Sound, the most splendid sheet of water for yachting purposes in the world, were the chief glory of the yachtsmen. To start

* First owned by Jacob Voorhis; present owner John R. Dickerson.

with a congenial party such as the yacht could comfortably accommodate, and go for a ten days' cruise to the eastward. These cruises, of course, were as frequent as the business engagements of the owner would permit, all through the yachting season, and long before the date of our last résumé, at the end of the summer of 1858, the pennant of the New York Yacht Club had been a common sight in every harbor from Glen Cove to Martha's Vineyard, the yachtsmen being always welcome visitors; and leaving always substantial pecuniary benefits behind them.

Continuing the history from the point where I left it, I may say that in 1859 another change was made in the system of measurement for allowance of time being by area of hull; length and breadth at the water-line only being taken into the account, and this method proved so satisfactory, that it was not changed until 1870, when in view of the arrival of the schooner *Cambria* to race for the *America's* Cup, the rule was changed to the cubical contents one, which, I think, was the fairest for all shapes of vessels, taking all things into consideration, of any that the club has ever adopted. The annual regatta was sailed this year over the old Elysian Fields course, and there was nothing particular about it except this change in the system of measurement from area of canvas to area of hull, a great improvement.

During the annual cruise, this year, on the 6th of August, Mr. Bennet matched his sloop *Rebecca* against the schooner *Restless* for $500 a side, to sail from Brenton's Reef Lighthouse, off the harbor of Newport, through the Sound to the Throggs Neck buoy, a distance of 154 miles. It was a very fine race, the wind being strong from the southwest, and the *Restless*, being by eighteen tons the larger vessel, beat the *Rebecca* twelve minutes. Two days later, August 8, the schooners *Favorite* and *Haze* sailed a match at New London, over a course 24 miles, and the *Favorite* won, and on August 10, the whole fleet had a race at Newport, the course being from off Port Adams, to a stake boat anchored sixteen miles southwest by south half south from the Brenton's Reef Lightship, and at this match, for the first time, two classes of sloops as well as two classes of schooners sailed.

The club had a fall regatta, this year, from off Owl's Head around the lightship, and at this there were three classes of sloops, and two of schooners. The race fixed originally for September 22 failed on that day from lack of wind, and was finally sailed September 26.

This seems to have been a yachting year, as on October 6 there was a match between the schooners *Gypsy* (148.94), *Favorite* (138), and *Zinga* (118.7), the course being from off Hart Island, to and around the buoy off Eaton's Neck and return, thirty-eight miles for $50 each, and the *Gypsy* won. The race was sailed under reefed sails, and the *Favorite* twisted her rudder head.

In 1860 the regular regatta was sailed, June 7, and, as in the last race, there were three classes of sloops and two of schooners, and they went over the old course from off the Elysian Fields. On August 2, of this year, 1860, the sloops *Julia* (85.3) and *Rebecca* (76.4) sailed a match twenty miles to windward, from Sandy Hook, for $250 a side. This was the first race over this course, since become historical, and the yachts sailed with housed topmasts by stipulation, and under jib and mainsail only. The *Julia* won by thirteen minutes.

On the annual cruise, this year, the fleet sailed a race at New Bedford, there being, as had now become the fashion, three classes of sloops and two of schooners, the *Julia* winning the champion prize for sloops.

In 1861, the New York Yacht Club had no regatta. This was in consequence of the breaking out of the war, but in 1862, they went at it again, with three classes of sloops, and three of schooners, divided into those of 800 square feet of area, from that to 1,300 for the second class, and over 1,300 for the first class of sloops. The schooners were up to 1,000 feet for the third class, between 1,000 and 1,500 for the second, and all over 1,500 feet for the first, and in this race the *Maria* for the first time sailed as a schooner. The race was over the old course, as was also that of 1863, in which year the annual regatta was sailed on June 11, and was a handicap, the first in the history of the club. It did not seem to give much satisfaction, as it was not repeated, but I fancy there was too much machinery about it, as the allowances were graded to fit any wind from a light breeze to a gale. It attracted, however, an entry of 9 schooners and 7 sloops, and was sailed over the old course.

June 3, 1864, at the regular regatta, the club went back to the old fashion of two classes of sloops and two of schooners, and they sailed over the old course; but the next year, on June 8, 1865, was sailed the first regular June regatta around the light ship from off Owl's Head, at which there was but one class of each rig, one of sloops and one of schooners, and they secured an entry of three of the single-masted vessels and six of the schooners. From this regatta ladies were excluded, it being thought that it would be too uncomfortable for them to go outside of the Hook, even in a well-appointed steamer. There was a strong breeze, and but three of the yachts were timed at the finish. The schooner *Magic* won here her first race. In order to compensate the ladies for their exclusion from the committee steamer on the day of the regatta, what was intended for a grand review was arranged for June 13, the place selected being the Horseshoe. Several similar attempts have been made in the history of the New York Yacht Club, which appears to have tried in turn almost every description of aquatic carnival, but all of them fifteen miles to windward from the light ship, the stipulation being that the tacks should be thirty minutes' duration, and that there should be no restriction as to canvas or number of crew, and no allowance of time. The wind was a fresh sailing breeze from southeast, and a thick fog shut down soon after the start. The *Magic*

SLOOP "REBECCA."*

have been more or less failures, the club never having taken kindly to reviews. On this occasion but thirteen yachts appeared, but the affair seemed so satisfactory to the committee, that in its report it expresses the hope that the review may be repeated each year, in which hope it was disappointed, for the ladies' privilege on the club steamer was restored to them, and there were no more reviews, at least not for many years.

June 13, 1865, five days later, there was a match sailed between the schooners *Magic* and *Josephine*, for $1,000 a side; got out to the mark all right and made the run back successfully. The *Josephine* failed to find the outer mark and lost the race.

Thus, it will be seen that, from year to year, each season brought something new, and this year was particularly fruitful of novelty, for on September 11, the first race ever sailed from Sandy Hook to Cape May was started, being a match between Mr. J. G. Bennett's schooner, *Henrietta*, 230 tons, and Mr. George A. Osgood's schooner, *Fleetwing*, 206.1 tons. The *Fleetwing* won by one and a half hours.

* First owned by Jas. Gordon Bennett; present owner G. P. Upham, Jr., Boston; now altered to a schooner.

Mr. Bennett was always ready for these matches, and October 16, of this same year, 1865, he sailed the *Henrietta* against the schooner *Palmer*, then owned by Mr. R. F. Loper. She entered at 194.22 tons, and the

Mr. Bennett made another match with the *Henrietta* this same fall. It was her first season, and he seems to have been inclined to race her for all she was worth. He wound up the season by sailing her against the schooner *Restless*, for $500 a side, the course being from Sands Point to the Bartlett Reef Lightship, off New London, and the *Henrietta* won by twenty minutes.

The annual regatta of 1866 was sailed, June 14, from Owls Head to and around the lightship, with the regulation single class of sloops and schooners, and nothing special occurred. The club seems now to have permanently abandoned the Elysian Fields course, and to have adopted Owls Head as the place of start and finish.

During the cruise, this year, a match was sailed betwen the schooners *Widgeon* and *Vesta*, on August 17, the stakes being $1,000 a side, and the course from off Fort Adams to and around the Block Island buoy, and return, which has come to be known as the regular Block Island course. This was a very close race, and the *Widgeon* won by one minute four seconds.

SCHOONER "SAPPHO."*

Henrietta seems to have been too large for her, as she beat her 21 minutes. The race was for $500 a side, as was also that between the *Henrietta* and *Fleetwing* on the previous month.

October 9, 1866, Mr. Bennett sailed the *Henrietta* against the *Vesta* in a match from Sandy Hook to the Cape May Lightship and return. There was a hard gale from the eastward, and both yachts were

* Built and owned by C. and R. Poillon ; purchased by W. H. Douglass ; present owner Prince Sciarra, Naples, Italy.

much damaged. The *Vesta* lost jib-boom, and the *Henrietta*, among other troubles, parted forestay, and had to lie to, for some hours, repairing damages. The *Henrietta* made the run down in 9h. 8m., and made the entire course in 30h. 6m., the *Vesta* beating her 56m., and winning the stakes which, as usual, were $500 a side.

This was a great year for match racing, and these matches were but the prelude to the greatest match ever sailed by yachts of any country, the great ocean race, which was started December 11, 1866.

Previous to this, however, on October 23, the schooners *Halcyon* (121) and *Vesta* (201), sailed a match for $250 a side, from Sands Point to the Bartlett Reef Lightship, the *Vesta* winning by nearly an hour. The *Vesta*, which at this time was owned by Mr. Pierre Lorillard, was sailed in a match twenty miles to windward from the Sandy Hook Lightship and return

SCHOONER "HALCYON." *

for a piece of plate against *L'Hirondelle*, afterwards the celebrated schooner *Dauntless*. She was entered in this race at 262.8 tons against the *Vesta*, 201, and as usual, size told in her favor, and she won. It was *L'Hirondelle's* first season, and she was owned by Mr. L. B. Bradford, from whom she was afterwards purchased by Mr. Bennett.

The great race from Sandy Hook across the ocean to the Needles, Isle of Wight, England, was the most remarkable contest ever entered into either on land or water. That vessels of the size of these schooners should cross the ocean at any time of year, was considered somewhat hazardous, but that they should cross in the dead of winter, added

* Original owners: W. Herbert, James E. Smith ; present owner Charles J. Paine, Boston.

immensely to the risk. Had they been especially prepared for an ocean voyage by having their spars reduced before starting, it would still have been considered something of a feat to have crossed the Atlantic in either of them in the month of December, but that they should start with racing spars and canvas to go across at racing speed, was something which all seamen would have considered imprudent. Then, too, the magnitude of the stake raced for, $90,000—a much more important amount then than now—added to the wonder of the undertaking, and finally, the passages made by all three of the yachts, all being little, if any, above the record made by the best appointed sailing packet ships, and below or about the average of steamer time in those days, placed the crowning glory on the enterprise, and I think, therefore, I am correct in calling this the most remarkable race of any kind on record.

Certainly, it was the most remarkable yacht race ever sailed, whether as regards the length and nature of the course, the season of the year, the amount of money involved, or the result, and therefore, I think I shall be justified in giving a more minute description of this race than I have been able to of any other within the limits of this article. For it was this race which lifted American yachting to a level with any in the world, and placed the New York club on an equality with the Royal Yacht Squadron of Great Britain.

"DREADNAUGHT."

As we have seen, American yacht owners had been yearly becoming more adventurous. The old club course had become too limited for them, and they had laid out a race track, a part of which was on the ocean. This had not satisfied them, and they had sailed races of hundreds of miles out on the ocean entirely, and on one occasion the track of a race had encircled Long Island.

Owners of the New York Yacht Club then, far more than now, were practical yachtsmen; that is, they sailed or knew how to sail, their own craft. Of course, some do this even now, but the proportion of experts among the New York Yacht Club owners is not, I think, as large as among the owners in the Atlantic, or Seawanhaka, Corinthian, or the Larchmont Clubs, and to go still further down in the scale of importance, the proportion of experts, that is, men who habitually sail the yachts they own, is greater in the Jersey City, New Jersey, and Knickerbocker Yacht Clubs, than in the others I have named.

In 1866, however, the Brooklyn club was but nine years old, the Jersey City but eight, the Boston but one, and the Atlantic Club just organized.

Practically, all the yachtsmen of this section belonged to the New York Yacht Club, and in those early days, few joined it who were not practical yachtsmen. This very brilliant feat of which I am writing, did much to attract to its rolls gentlemen from all the professions of life, and the jurist, the lawyer, the doctor, the merchant, esteemed it an honor to belong to this famous organization, and its

had been ample time for consideration. The probability is, that inasmuch as the two gentlemen who first made the match were enthusiastic yachtsmen and keen sportsmen, they needed no other inspiration than their own love of sport, and had no other. I give the original agreement *verbatim.*

"George and Franklin Osgood bet Pierre Lorillard, Jr., and others, $30,000 that the *Fleetwing* can beat the *Vesta* to the Needles, on the coast of England, yachts to start from Sandy Hook on the second Tuesday in December, 1866, to sail

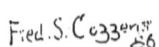

SCHOONER "COMET."

members were nearly doubled within the year.

It is said that this ocean match was originally made as an after-dinner inspiration over the wine; but although this might have been true as to two of the gentlemen engaged in it, it certainly was not as to the third, for he came in subsequently, and after there

according to the rules of the New York Yacht Club, waiving allowance of time. The sails to be carried are mainsail, foresail, jib, flying jib, jib topsail, fore and main gaff topsails, storm staysail and trysail."

This shows that in the process of evolution, schooners had come to a fore topmast and to a flying jib boom. At first they only had a little stump of a bowsprit and a short main topmast—more a flag-staff than anything else—on which was hoisted a sort of square topsail with a yard on it, sent up from the deck flying. The modern gaff topsail, now in universal use on fore and aft vessels was not introduced until some years after the organization of the New York Yacht Club.

To return to my story of this great race, in which there is ample material for a history by itself, and which the limitations of space forbid my more than merely glancing at. As soon as Mr. Bennett heard of this match having been made, he signified his desire to take a part in it, and, after some consideration, the other gentlemen consented, an article being added to the agreement as follows :

"The yacht *Henrietta* enters the above race, by paying $30,000 subscription by members of the New York Yacht Club; any minor points not embraced in the above, that cannot be settled by Messrs. Osgood, Lorillard and Bennett, shall be decided as follows : Each shall choose an umpire ; the umpires chosen in case of a disagreement to choose two others. Twenty per cent. of the money to be deposited with Mr. Leonard W. Jerome, on the 3d of November, the balance to be deposited on the first Tuesday in December—play or pay.

Signed by *J. G. Bennett, Jr.,*
Franklin Osgood,
George A. Osgood,
Pierre Lorillard, Jr.

December 5, 1866."

There was a supplementary agreement which provided that neither yacht was to take a channel pilot from this city, and that, in addition to the sails previously named, each yacht might carry a square sail. The third agreement provided

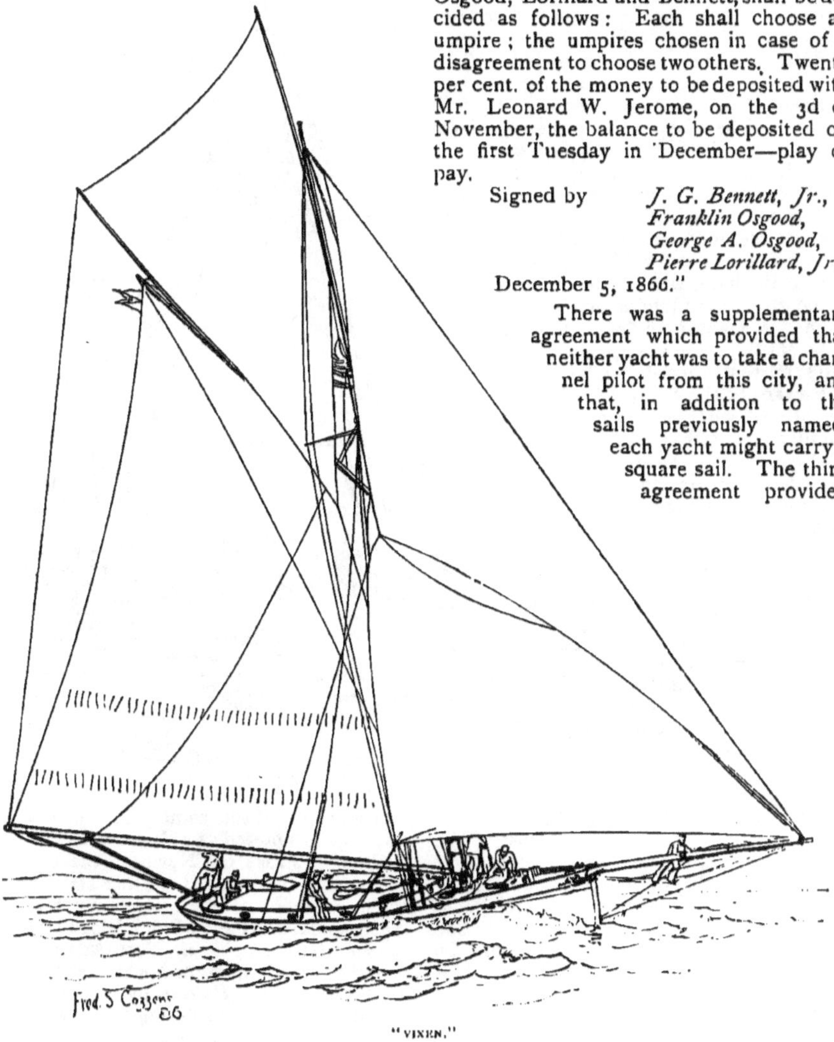

"VIXEN."

that each yacht might shift during the race everything but ballast, and that the forty-eight hour rule should be waived (that is, they could trim ship up to the very moment of starting). The race to end when the lighthouse on the west end of the Isle of Wight appears abeam, with the yacht on the true channel course, yachts to start on Thursday, December 11, at 1 o'clock P.M., blow high or low. Boats to be started by H. S. Fearing.

I have no space to dwell on their passages, although the logs of all three are before me. We know that they had a fine, fair start, and the result shows how wonderfully well they were navigated. The *Henrietta* won, having sailed 3,106 miles in thirteen days, twenty-one hours, fifty-five minutes. The *Fleetwing* was second, having sailed 3,135 miles, in fourteen days, six hours, ten minutes. The *Vesta* (fastest of the three) came last, having sailed 3,144 miles in fourteen days, six hours, fifty minutes. She was the only center-board boat, and on the day before their getting in with the land, was ahead of both of the others; a blunder on the part of her navigator in not allowing sufficiently for the strength of Runnell's current, caused her to fall in to leeward of the Scilly Islands with a southerly wind, and a more cruel blunder of her channel pilot caused her to run past her port in the channel and lost her the second place, showing once more that "the race is not always to the swift."

The only accident happened to the *Fleetwing*, while scudding before a hard gale, December 19, under a double-reefed foresail and fore staysail. At nine o'clock in the evening, she took a sea aboard which washed six of her crew out of the cockpit, and they were lost. The boat was then obliged to lay to for five hours, under her double-reefed foresail.

It was in 1867 that the schooner yacht *Sappho* made her first appearance. She was built in Brooklyn, by the Poillons, on speculation; a deep keel vessel, with finer lines than had been the fashion previous to that, and her builders confidently expected that she would prove faster than any yacht afloat. She did so prove afterwards, but her early career was not promising. She sailed her first race off New London, August 7, 1867, a match of thirty-five miles for a cup offered by the commodore of the club, in which five sloops and seven schooners started. There was a thick fog, and some of the yachts did not return until after midnight. The schooner *Eva* was the only one that made the race inside of the time limit.

"VESTA." *

August 10, of this same year, the *Sappho* was again entered by her builder, Mr. Poillon, in a race off Newport, the course being from Brenton's Reef to a stake boat anchored about a mile east by north from the lighthouse on Sandy Point, Block Island, returning to a point off Port Adams, the race to be made in eight hours. She came in second to the *Palmer* by two minutes actual time, and, considering the difference

* Originally owned by Mr. Pierre Lorillard, and now by Mr. Fred. F. Ayers.

THE HISTORY OF AMERICAN YACHTING.

"WANDERER."

in size, this was a bad beat for the new schooner, from which so much had been expected. Here are the dimensions of the two boats:

Sappho, keel; 274.4 tons; 3146.0 feet area.

Palmer, center-board; 294.2 tons; 2371.9 feet area.

1868 was notable as being the year when the club established itself at Clifton, S. I., and for the first time started its annual regatta from there on June 18, 1868. There were four sloops and eight schooners started, and the affair failed from lack of wind, and next day only two sloops, the *Gussie* and *White Wing*, came to the line with the schooners *Magic, Idler, Silvie* and *Rambler*. As has generally happened on days when postponed races have been sailed, there was a cracking breeze,

the *White Wing* was disabled, and the *Magic* took the schooner prize. A famous race was sailed July 15, of this year, between the schooners *Magic* and *Pauline;* the owner of the *Magic* betting $3,000 to $2,500! the course being the regular one around the lightship; the *Magic*, 112.5 tons, allowing the *Pauline*, 81.2 tons, seven minutes. They started from an anchorage, as was the custom for some years later, and the *Pauline* led her larger competitor all the way around the course, and beat her, finally, thirty minutes fifteen seconds, actual time. This was the worst beating the *Magic*, a wonderfully smart boat, ever received, and showed most conclusively the uncertainties of yacht racing. The wind at the first was variable, but at the Hook they got a fresh breeze from southeast, at least the *Pauline* did, and away she went, getting clear out to the lightship before the *Magic*, inside of the Hook, got the breeze at all.

During the annual cruise, this year, there were some fine matches, but nothing especially worthy of note. Mr. Pierre Lorillard gave a cup at New London. Mr. Thomas Durant, at that time owner of the schooner *Idler*, gave one at Newport, and there was an ocean sweepstakes from Clarks Point, off New Bedford, twenty miles to sea and return, and to be made in five hours. Of course, it was not made in that time. The Poillons, meanwhile, not having been able to secure a purchaser for the *Sappho*, had sent her to England for sale, and she sailed a match around the Isle of Wight. It was a sweepstakes, £2 entrance money; the race to be made in nine hours. The *Sappho* entered at 310 tons. Cutters were to have two thirds of their tonnage added. There were no square sails allowed, but in fore and aft canvas there was no limit. No greater amount of time than twenty minutes to be allowed in any event.

Evidently, the big Yankee schooner did not frighten John Bull to any great extent, for the cutter *Oimara* undertook to sail the *Sappho* on even terms, while, as respected English yachts, she was to have two-thirds of her tonnage added, and was to be classed at 275 tons. The other yachts were the cutter *Condor* (215), and the schooners *Cambria* (193) and *Aline* (212). The *Cambria* was owned by Mr. Ashbury, and the *Aline* by Mr., now Sir Richard, Sutton, who came here last summer with the cutter *Genesta*. The above were the measurements sailed under, the real measurements of the *Oimara* and *Condor* being 165 and

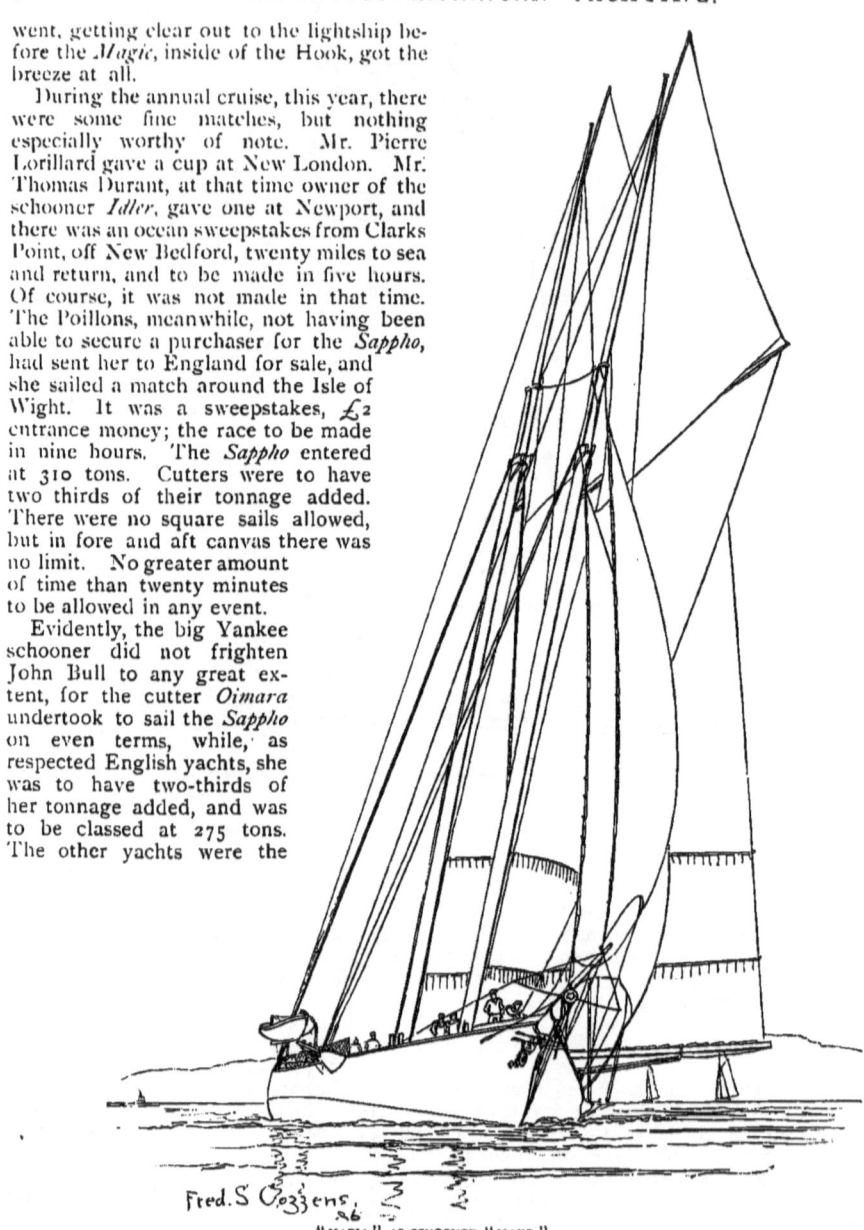

"MARIA" AS SCHOONER "MAUD."

129 tons respectively. The allowances were: *Sappho* allows *Oimara* .00; *Condor*, 9.12; *Aline*, 9.12; *Cambria*, 11.55. *Oimara*, allows *Condor* 6.16; *Aline*, 6.16; *Cambria*, 8.59. There was a fine breeze northwest, and the yachts came in: The *Cambria*, 6.17.50; *Aline*, 6.19.55; *Oimara*, 6.23.10; *Condor*, 6.25.00; *Sappho*, 7.58.00.

The *Sappho* lost jib boom off Ventnor, and about half way over the course. She ought not to have started at all. The gentleman in charge of her, a good navigator and thorough business man, was not a yacht-racing skipper, and this defeat settled all chance of selling the yacht, and she had to return to this country unsold. After her return, she was purchased by Mr. W. P. Douglass, recently the vice-commodore of the New York club, and under the direction of the late Captain Robert Fish, she was hipped out, and began at once a most successful career.

Marine architects differ in opinion as to the value of the alteration. The builders of the *Sappho*, to this day, are of the opinion that she was as fast before as after the alteration, and that her excess of sail-carrying power, resulting

was remarkable as having had for prizes cups presented by James G. Bennett, Jr., then the vice-commodore of the club. When Mr. Bennett first became a member of the New York Yacht Club, there was a strong prejudice against him, on the part of some of the older and more aristocratic

"GRACIE."*

from the hipping, was more than counterbalanced by the increased resistance. The fact, however, is patent, she was a failure before, and a grand success after the alteration.

The autumn regatta of the New York club for 1868 was sailed September 22, and

members. He was considered by them a sort of *parvenu*, and it was the influence of this feeling that ruled his yacht out when she had won the race around Long Island rather than because she had deviated from the course by coming through Plum Gut; for, as is well known, at certain times of

* First owner William Voorhis, then Wm. Krebs, next John P. Waller; present owner Joseph P. Earle.

the tide, nothing can be gained by going through this passage.

Gradually, however, Mr. Bennett's true sportsmanlike spirit found proper recognition in the club, and on his return from the great ocean race — he having pluckily gone out and returned on his yacht — he was unanimously elected vice-commodore, and, as stated, at this regatta he presented the prizes. The race was over the regular course, but to be sailed without time allowance, and to be made in seven hours. The starters were of schooners: Mr. R. F. Loper's *Palmer* (194.2), Mr. E. Dodge's *Silvie* (106.2), Mr. H. G. Stebbins' *Phantom* (123.3). Sloops:

"ENCHANTRESS."

Mr. John Voorhis' *Addie V.* (44.8), Mr. William Voorhis' *Gracie* (54.5), Mr. Sheppard Homans' *White Wing* (53.1), Mr. John B. Herreshoff's *Sadie* (42.1). I give these names and the names of the owners, because nearly all these yachts are still in commission, all of them, in fact, except the *Silvie*; but all have undergone extensive alterations since that time. The *Palmer* has been rebuilt and raised; the *Phantom* is, I think, about the same as then, except that she has each season been overhauled and kept in prime condition; the *Addie V.* and *Gracie* have been rebuilt and enlarged, the present *Gracie* being nearly twice as large as she was then. The *White Wing* has gone through many changes. She was sold out of the club, and used down at Greenport, L. I., as a bunker boat. Then, in 1878, I think, she was purchased by Mr. A. Perry Bilven, extensively repaired and rebuilt, and called the *Ada*, but there being some custom house informality about the change, she resumed her old name and is still running, enrolled in the Brooklyn and Hull clubs. The *Sadie* has been enlarged; a second mast added, and she is now the schooner *Lotus*. In this match for Mr. Bennett's cups, the wind was fresh from southeast, and the winners were the schooner *Phantom* and the sloop *Addie V*.

At the twenty-third annual regatta — only the June matches are enumerated — there seems to have been a spar-breaking breeze, and the number of lame ducks which limped back to the anchorage in the afternoon was considerable. The match was sailed June 10, 1869, and the starters were five schooners, seven sloops over 25 tons, and three sloops under 25 tons. Summing up the result, the *Phantom* carried away mainmast head, between the Hook and lightship, bound out. Schooner *Silvie* lost flying jib-boom, and *Palmer* carried away fore topmast. The winners were: schooner *Idler*, sloops *Sadie* in the first, and *White Cap* in the second class.

July 10, 1869, there was a match race over the regular club course, between Mr. James M. Banker's schooner, *Rambler* (164.4), and Mr. Franklin Osgood's schooner *Magic* (97.17), Mr. Banker betting $1,000 to $500 that his boat, without time allowance, would win; in which, as the result showed, he was slightly mistaken, for the little *Magic* led all around the course. The start was from an anchorage with a light southerly breeze, which freshened after the yachts had passed through the Narrows, backing to the southeast, and was quite fresh between the Hook and lightship, at which mark the *Magic* was a long distance ahead; but as most frequently happens, the leading yacht lost the wind on drawing in to the Bay on the return, and the *Rambler*, retaining the ocean breeze, came nearly up with her; then the little boat drew away again and went in an easy winner, the *Rambler's* discomfiture being made more complete by her hanging on the rocks off Fort Lafayette for two and a half minutes.

This schooner *Rambler* was not the present schooner of that name, as she was not built until 1871. She was built also for Mr. James M. Banker, by Mr. E. P. Beckwith, at New London, and afterwards sold to Mr. John M. Forbes, and by him to the late commodore William H. Thomas, who had her very much lengthened, in 1876, by Mr. Downing, of South Brooklyn.

There was a race for three cups, this year, from New London to Newport, the classes being over 120 tons, under 120 tons for schooners, and the third cup for all sloops, no allowance of time. The wind was moderate from S. S. W., and the winners were: schooners *Rambler* and *Magic*, and the sloop *Gracie*.

August 11, 1869, there was a race without allowance of time, over the regular Block Island course, for three cups, two for schooners and one for sloops. The lucky boats were the schooners *Phantom* and *Eva*, and the sloop *Gracie*.

We now come to the great American yachting year, 1870, the most important in the history of the sport of any that has occurred since the introduction of yachting in this country, in 1844. It began on the opposite side of the Atlantic. After Mr. Douglass had completed the alterations in the schooner *Sappho*, of which I have already written, he started in her for England, determined that she should retrieve her record in that country if it was possible to do so. Mr. James Ashbury, the owner of the schooner *Cambria*, gladly acceded to the desire of Mr. Douglass for a race, and a match was made for three races, each for a fifty-guinea cup. Two of them were to be sixty miles to windward, and the third a triangle, with sides twenty miles long.

The articles of agreement were most elaborate, and were probably drawn up by Mr. Dixon Kemp, who acted for Mr. Ashbury on this occasion, and who afterwards accompanied that gentleman, when he came to this country, in the *Cambria* during this same year. The articles were signed by Mr. Kemp for Mr. Ashbury, and by Mr. J. D. Lee for Mr. Douglass.

The *Sappho* went in at her whole tonnage, 310 tons, and the *Cambria* at 199 tons. Evidently, Mr. Ashbury did not have that healthy respect for the *Sappho* which he came to have afterwards, for he sailed this match without allowance of time, and of course there could be but one result. In the first race, after getting over about forty miles of the course, the *Cambria* found herself so far astern that she bore up. This was on May 10. On May 14 the second trial came on, and the judges fixed the course from the Nab to the Cherbourg breakwater, sixty-six miles south-west. The wind was west-south-west, and Mr. Ashbury, or more properly Mr. Dixon Kemp, who acted as his adviser, protested that it was not dead to windward, and the judges "sat down" on Mr. Ashbury very properly, and told him it was near enough to dead to windward, and as fair for one boat as for another; also, that it was the best they could do for him. Whereupon he refused to start, and the *Sappho* went over the course alone.

The third race came off May 17, the courses being west-south-west, south-east half east, and north three-quarters east. The times at the first mark were: *Sappho*, 1h. 7m. 35s.; *Cambria*, 1h. 11m. 14s. At the second mark: *Sappho*, 4h. 20m.; *Cambria*, 6h., and the time at the finish need not be given. Of course, the American yacht was declared the winner of all three cups.

It was on June 14, 1870, that the twenty-fourth annual regatta was sailed, and there were one class of schooners and two of sloops. There were eleven schooners took part in the match, but very few sloops. Just then the schooner was the popular rig, as it will possibly be again, but it is hardly probable; the men of wealth at present who would have built schooners in the olden time, will now build steamers as a handier cruising vessel. Nearly all of these eleven schooners are still among the yachting fleet; they were the *Madgie, Magic, Fleetwing, Tidal Wave, Madeleine, Alarm, Silvie, Palmer, Phantom, Alice,* and *Idler*. The latter was the winner, and the successful sloops were the *Sadie* and *White Cap*.

The next event, in this yachting year, was the ocean match, from Gaunt Head, Ireland, to the Sandy Hook Lightship, by the schooners *Cambria* and *Dauntless*, for a cup of £250. They entered under the following measurements:

Name.	Owner.	Tons Measurement. N. Y. Y. C.	R. T. Y. C.
Cambria,	James Ashbury,	227.5	188
Dauntless,	James Gordon Bennett, Jr.,	268.0	321

They started July 4, and on arrival, although the lightship was the terminal point of the race, the official time was taken as the yachts passed the buoy off Sandy Hook. The *Cambria* arrived July 27, at 3.30 P.M.; the *Dauntless*, July 27, at 4.47 P.M., a difference of 1 hour, 17 minutes. They were both navigated by old merchant captains, the *Cambria* having Captain Tannock, who had commanded ships in the trade between Liverpool and Quebec and Montreal; and the *Dauntless* had Captain Samuels, of *Dreadnaught* fame, who, in the *Henrietta*, had won the great ocean race from Sandy Hook to Cowes. On board of the *Dauntless*, also, were "Old Dick Brown," who was in the *America* when she won her great race in 1851, Captain Martin Lyons, a Sandy Hook pilot of great experience, and Mr. Bennett was also on board; and it is just possible, and altogether probable, that her defeat was due to this excess of talent on board of her. The official record of the two passages shows that the *Cambria* was navigated the best. She sailed 2,917 miles in 23 days, 5 hours, 17 minutes, 5 seconds, and the *Dauntless* sailed 2,963 miles in 23 days, 7 hours. The excess in distance, 46 miles, is more than equivalent for the difference in time, 1 hour, 42 minutes, 55 seconds.

We come now to the first great race in this country for the *America's* Cup, which had been made by the owners of the *America* a perpetual international challenge prize, and as such had been held in trust by the New York Yacht Club; and as this race was followed by many others, in which the British schooner *Cambria* participated, I may well leave them until a future chapter, merely at this point recording my opinion that this schooner was the smartest of any of the vessels of that rig that have come here for the *America's* Cup. Her model is in the model room of the New York Yacht Club, and in that, the largest collection of models in the world, there is none that surpasses it in gracefulness of outline. She was beaten because she was clumsily rigged and canvased. British sailmakers have made an immense advance in the making of sails for yachts within the past fifteen years, and I think I am correct in saying that with a modern Lapthorne suit of canvas, the *Cambria*, in 1870, would have carried home the *America's* Cup.

THE HISTORY OF AMERICAN YACHTING.

BY CAPTAIN R. F. COFFIN,

Author of "OLD SAILOR YARNS," "THE AMERICA'S CUP," etc., etc.

III.

THE INTERNATIONAL PERIOD.

AFTER the year 1870, yachting in this country broadened out — became diffused. For the first twenty years, it had been almost wholly confined to New York and its vicinity, and down to the end of the year 1869, twenty-five years after the parent club had been organized, there were not many more than a dozen yacht clubs in the whole country. There were, however, very many small sailing craft held by individual owners, and from this time on, these, in various sections, have banded themselves together, and formed yacht clubs, with very beneficial results. Designing, and the draughting of sail and spar plans, which, previous to this, had been confined to professionals, builders, riggers, and sailmakers, has been studied by the young men of the clubs, and the result has been a great improvement in the appearance, as well as the performance, of American yachts. Young men whose privilege it has been to travel, have studied the methods of British yachtsmen, and the designs of British yachts, and have returned to this country with enlarged ideas as to the future possibilities of the sport, and the result is a type of yacht suited to our shallow and generally smooth waters, which combines in her design and rig some of the best features of the British yacht. The old notion that a body could be moved over the water easier than through it has been found to be untenable, and we are building now, vessels of more moderate beam and of increased depth. In ballasting the yacht, also, there has been an immense improvement. Formerly, it was not thought that anything more expensive than scrap-iron or paving-stones could be afforded for ballast, and it was a great advance when we got to moulding the iron to fit the frames, and in this manner lowering the ballast and securing increased stability with less weight than before. The substitution of lead for iron was another advance, and for sharp-bottom boats the placing of the lead outside is an improvement. Our old-fashioned, flat-bottom, light-draught boats, however, do not need it there, and in some instances it has proved a detriment, and has had to be removed.

In rigging and in canvas, we have been constantly improving; wire has entirely superseded rope for standing rigging, and the bringing of the head stay to the knight heads with runners from the mast-head aft, has given a stability to the mast which prevents the canvas from getting out of shape in strong breezes. This, of course, necessitates a double-head sail instead of the one large jib formerly used, and although there is in this substitution a loss of propelling power in moderate breezes, a defect of the old rig is cured by this substitution, and the yacht is handier in a reefing breeze. Formerly, when the wind increased so that the mainsail of the sloop had to be reefed, there was a difficulty in reducing the forward canvas. A reef was clumsy in a jib; a bonnet — for a yacht — nine-tenths of whose service is in whole sail breezes, was scarcely to be thought of, and a "bob jib" was an abomination. With the double-head sail, the difficulty is obviated, and generally, the small jib can be carried in any breeze to which the usual service of a yacht exposes her; and at all events, she can always carry the fore staysail. A mistake made by sailing masters

SLOOP "BIANCA."

at the first introduction of the double-head sails, was that, in racing, they took in the jib first. This should never be done, as long as it can be carried; as there is but slight propelling power in the staysail.

The advantage, however, of the large jib is so apparent, that some of the yachts have their forestays fitted so that they can be come up with at will, and the big jib can be used, if necessary, in races; while, for ordinary sailing or cruising, the handier double sails are used. From year to year, however, we have been improving, and where we formerly used ordinary canvas, such as was made for coasters generally, now, for a yacht of any pretensions, the canvas is manufactured especially for her, and of narrow cloth.

It was in 1870, that the Eastern Yacht Club — now one of the most important in the country — was organized. The Dorchester and the Manhattan also came into existence this year, and were followed next year by the Seawanhaka and the New Jersey, the latter securing the old quarters of the New York club at the Elysian Fields. From this time on, clubs have multiplied to an enormous extent, and especially in the New England States. All the lake ports have their yacht clubs, and there are three or four on the Pacific Coast. The South, too, has its yacht clubs, some of them very thriving organizations. In Philadelphia, Baltimore, Charleston, Savannah, St. Augustine, Mobile, and New Orleans, yachting is active. Still, these organizations are all comparatively young, the Quaker City, at Philadelphia, having been born in 1876, and the Mobile not until 1883.

So, then, the interest of the public at large still centered around the operations of the New York club, and especially in its defense of the *America's* Cup. We all remember the furore of excitement that there was here, last season, when the cutter *Genesta* came, and the *Cambria*, when she came for it, aroused quite as much interest, if not more. The mosquitos at Sandy Hook fed on a small army of reporters for a week before she arrived; and Mr. Ashbury's movements were chronicled in the daily papers, as those of Sir Richard Sutton's were last summer; and on the day of the race,

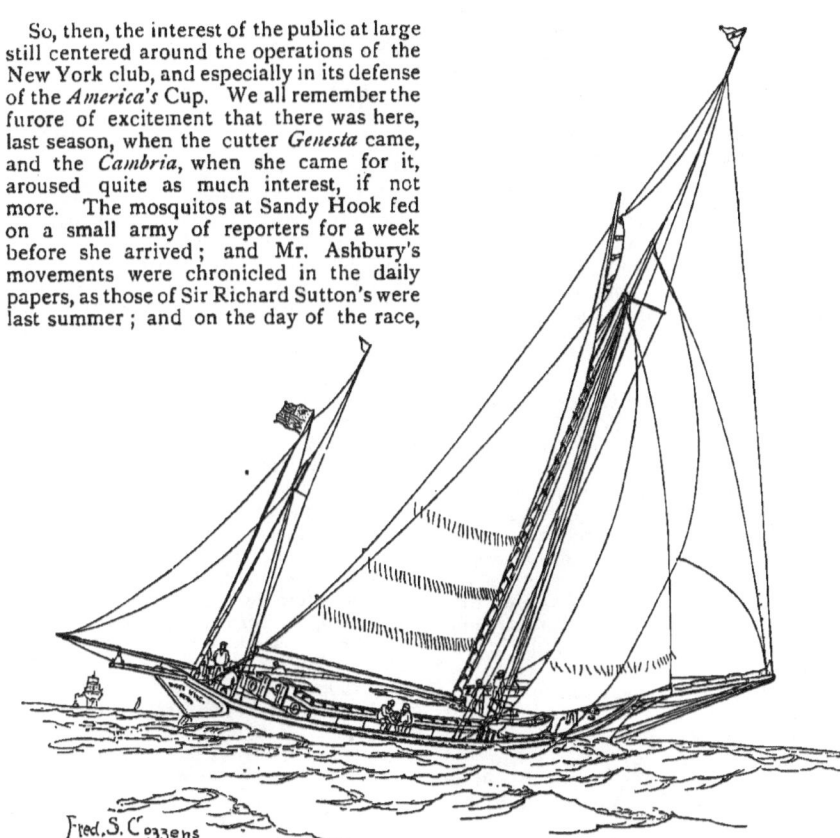

YAWL "WHITE WING."[1]

which was August 8, 1870, nearly all downtown business was suspended, and Broad and New streets were well nigh deserted, for all New York was upon the water; and if at the Light-ship and the finish there were not as many excursion steamers as there were last summer, it was because in 1870 there were not as many in this harbor as now.

The *Cambria* had to sail for the cup as the *America* first sailed for it — against the whole fleet. Mr. Ashbury protested against this, claiming that the word "match" in the deed of gift meant a duel between two vessels only, and that the New York club was bound to put a single representative vessel against the *Cambria*; but the club, by a vote of 18 to 1 (only yacht owners can vote), decided that

inasmuch as the *America* was obliged to sail against the whole fleet in order to win the cup, so all subsequent competitors for it must do the same. In this race, however, none but schooners started, and of these, counting in the *Cambria*, there were twenty-five, but only fifteen finished. The little *Magic* won on elapsed, as well as corrected time; and on elapsed time, the *Dauntless* was second. On corrected time, however, the *Idler* was second, *Silvie* third, and *America* fourth. The *Cambria* was tenth on corrected time. Mr. Ashbury, with the *Cambria*, accompanied the New York Yacht Club on its annual cruise, and there was a series of races at Newport which are memorable.

The whole fleet raced from New London to Newport, telegraphing on and obtaining

[1] Original owners Wm. and John Jacob Astor, Shepard Homans and a dozen others; present owner C. H. Bliven.

the consent of some gentlemen of that place to await the arrival of the yachts and time them. The *Tidal Wave* was the winner, with the *Cambria* well in front.

August 16, 1870, began, at Newport, a series of races the most brilliant and interesting in the history of American yachting.

The race upon this day was for two cups of 50 guineas each for schooners and sloops, presented by Mr. Ashbury, and a cup of the same value, presented by members of the New York Yacht Club, for the second schooner in on time allowance. This was to afford the *Cambria* an opportunity of sailing in the match. She could not go for her own cup, but in case she came first, she could change places with the second boat and take the prize. The course was the regular one, from off Fort Adams, around the Block Island buoy and return, and there started thirteen schooners and four sloops. No one cared much for the sloops in those days, and perchance should a schooner again come for the *America's* Cup, the two-masted vessels would again come into fashion. I think, however, that this is doubtful, for I believe that as racing craft, their day has passed on both sides of the Atlantic.

The little *Magic* again came in a victor, with the *Cambria* only 26 seconds behind her in actual time. So the *Magic* took the Ashbury Cup and the *Cambria* had the subscription cup, while the sloop *Gracie* took the Ashbury Cup for yachts of her rig.

Next day, August 17, 1870, the schooners *Cambria* and *Palmer* sailed a race over the Block Island course for a 50-guinea cup. This was Mr. Ashbury's standing wager, and for this, he was willing to sail any or all of the schooners of the club in succession. Their ratings for this match were:

YACHT.	OWNER.	AREA.
Cambria	James Ashbury	2,105.8 sq. ft.
Palmer	Rutherford Stuyvesant	2,371.9 "

The race began with a moderate breeze from south-west, backing to south-south-west and freshening to a good whole sail breeze. Evidently, the *Palmer* was the better boat, and she won by nearly seven minutes.

Next day, the *Cambria's* match with the *Idler* came off, and it was the only one of all her matches which she won. As I sailed on the *Cambria*, on this race, I am perhaps better competent to tell the reason of the *Idler's* defeat than most of those who have written about it. The two yachts scored as follows:

YACHT.	OWNER.	AREA.
Cambria	James Ashbury	2,105.8 sq. ft.
Idler	Thomas Durant	1,934.6 "

The club record of this race says that the wind was fresh from south-west. The direction is correct, but the expression of force is misleading, as the wind was just a fair whole sail breeze. During the previous day and night there had been a good breeze from south-south-west, and the yachts on the starboard tack, heading about south, encountered a rather troublesome head sea. The *Cambria* went off with the lead, the *Idler* in her wake, as both reached to the southward. As from time to time I allowed my head to get above the *Cambria's* rail to glance at the boat in her wake, I saw plainly that she was gradually eating across our wake and gaining position on our weather quarter. Suddenly, about a half hour after leaving the Brenton's Reef Light-ship, the *Idler* tacked, and it became a serious question with us whether we should allow her to go off alone. The plain rules of racing required that we should go around after her, but the southern tack was so manifestly the best, the westerly tide being on our lee bow, that we continued on. We were more than satisfied with this course when, upon tacking later on, the yacht's head came up to west-by-south, and sometimes west-south-west, and we weathered the *Idler* very neatly, and fetched the mark; and at the finish, the *Cambria* had the race by nearly 8m., corrected time.

But then I learned that the reason for the *Idler's* tacking and leaving us was, that the plate to which the bob-stay sets up had drawn out of the stem, and she could no longer head the sea, without danger of losing bowsprit and masts. On the port tack, the sea was more abeam, and to a strap through the hawser-holes a tackle was got to the bowsprit end, and the course was completed. I am certain, however, that the *Idler* lost more than eight minutes, and therefore, but for this accident, she would have won. It must be remembered that at this time neither *Idler* nor *Palmer* were what they were made to be later on, very extensive alterations having been made in both yachts, under the superintendence of Mr. Henry Steers.

At this time the late Henry G. Stebbins was the commodore of the club, and Mr. Wm. P. Douglass was the vice-commodore, Mr. J. G. Bennett was the rear commodore, and before the fleet left Newport to proceed to the eastward, in continuation of its cruise, Mr. Bennett offered a cup to be raced for on or after September 5, on the return of the fleet to Newport, by schooners solely, the course to be over what has since been known as the long Newport Course, from the Light-ship on Brenton's Reef to and around the Block Island buoy; thence, to the Sow and Pigs' Light-ship, and back to the place of departure, without allowance of time ; each yacht to subscribe $25, to be invested in a cup for the second schooner, or for the *Dauntless* should she be first ; this second prize to be determined by allowance of time. At the same time, Mr. Douglass offered a prize for schooners, to be sailed on the 6th of September, from the Light-ship on Brentons' Reef to the Block Island buoy and return, without allowance of time, with $25 subscripton, as in the Bennett race for second schooner, with allowance of time to be given to the *Sappho*, if first.

At the same time, Mr. Ashbury offered a prize for schooners and one also for sloops, to be sailed September 7, from Bateman's Point around the Block Island buoy and return, with time allowance ; but if no sloops enter, then the first schooner to take both prizes ; and Mr. Rutherford Stuyvesant offered a prize for the *Cambria*, if she was either first or second schooner on allowance of time, thus making a race for her, as she could not

SLOOP "FANNY."

SCHOONER "MAGIC."[1]

win her own cup. Having thus arranged for plenty of sport on the return, the club went on to the eastward, sailing first to New Bedford; and I presume that the entrance of the fleet into that old whaling port will never be forgotten by those who witnessed it; the *Dauntless* and *Cambria* coming in nearly side by side, before a fresh breeze, carrying square foresails and fore top-sails, water-sails and ring-tail even, until they rounded to at the anchorage, and then letting everything come down by the run.

At Oak Bluffs, Martha's Vineyard, which was the next stopping place, the fleet was caught in a north-east gale, which came on suddenly in the night, and a few of the yachts were blown ashore. The rest, and among them the *Cambria*, succeeded in getting under way, and made a harbor at Edgartown.

After the return of the fleet to Newport, the first of the series of arranged races noted above, namely, that for the Bennett Cup, was sailed September 8, with a fresh breeze, south-east, the starters being the *Cambria, Sappho, Palmer, Vesta, Tidal Wave, Idler, Madeleine, Halcyon, Phantom* and *Madgie*. The only two timed on the conclusion of the sixty-four-mile course

[1] Original owner Franklin Osgood; afterwards J. Lester Wallack and Rufus Hatch. She had formerly been called the *Madgie*. She was built in Philadelphia. Present owner Chas. G. Weld, of Boston.

were the *Palmer* at 6h. 34m., P.M., and the *Cambria*, 6h. 38m., P.M. The *Sappho* lost main topmast and split mainsail. The *Cambria* took the subscription cup. The other races of this series were postponed until after arrival in New York, but on September 9, the *Cambria*, *Phantom* and *Madeleine* sailed a match over the regular Block Island course for a 50-guinea cup; and with a fresh breeze south-south-west, the *Phantom* beat the *Cambria* 23m. 53s.; and the *Madeleine*, which came second, beat the British schooner 9m. 43s., after having been crippled by carrying away one of her bowsprit shrouds at the beginning of the race. This was the most crushing defeat which the *Cambria* encountered while in this country. The citizens of Newport now, in return for nearly a month's patronage by so fine a fleet as this, and in honor of Mr. Ashbury, offered a cup valued at $500, with a subscription club cup for second schooner; and this race, one of the most notable in the club's history, was sailed September 11, over the regular club course. There were eleven schooners started, but only four — the *Palmer*, *Phantom*, *Dauntless* and *Cambria* — were timed on the return. The start was made in a light air from south-west, which increased afterwards to a fair sailing breeze, and the yachts beat down to the buoy and rounded it. Just after getting all fancy kites aloft for the run back, the wind shifted in a hard squall to north-east, settling into that point, after the squall had passed, a reefing breeze with rain. Every yacht in the fleet save the *Cambria* met with more or less mishap, the *Dauntless* losing fore topmast.

The yachts arrived in the order given above, the *Phantom* (at that time flag-ship) taking the City Cup, and the *Palmer*, "scooping," the subscription prize. The *Cambria* was a long way astern of the *Dauntless*, and the rest of the fleet did not arrive until long after night-fall.

On the return of the fleet to this city, the racing was resumed, the New York gentlemen apparently determined to give Mr. Ashbury all the sport he desired, and to send the *Cambria* home with her locker full of cups. The season was getting advanced, and the honored visitor was becoming a trifle impatient, desiring to get home before the storms of the winter came on; so, on September 28, the Newport prizes, left over, were sailed for at the same time; viz., the cup offered by Mr. Douglass without allowance with Mr. Ashbury's cup for sloops and schooners, both cups to go to best schooner if no sloops started, and the cup offered by Mr. Rutherford Stuyvesant for the *Cambria* if she was either first or second; and the course was from buoy No. 5½ off the Point of Sandy Hook, twenty miles to windward and return. In this race, the *Cambria* was nowhere. The *Dauntless* won the Douglass Cup, the *Tidal Wave*, the Ashbury Cups, and the *Madeleine* the Stuyvesant Cup.

October 13, the great match of the *Sappho* and *Cambria* was sailed twenty miles to leeward from the Sandy Hook Light-ship and return for a 50-guinea cup, the race to be made in five and a half hours. The *Sappho* beat the *Cambria*, 50m. 50s. in this race, the wind being strong from north-west. The race was not made inside of the stipulated time, however, and no prize was given.

Next day, the *Dauntless* and *Cambria* raced for 50 guineas, twenty miles to windward from buoy No. 5 off the Hook, and return, and the *Dauntless* won by 12m. 30s. actual, and 7m. 18s. corrected time.

This satisfied Mr. Ashbury, and he soon after took the steamer for home; Captain Tannock taking the yacht across, and one more grand ocean race, between the *Sappho* and *Dauntless*, in which the *Sappho* won by 12m. 45s., earning the title of "Queen of the Seas," concluded the yachting of this racing year.

During the ensuing winter, as it was tolerably certain that Mr. Ashbury was to return with a new yacht, the measurement of the club was changed, so that the skimming dish should have rather the best of it, in competing with the deep keel yachts, and cubical contents were substituted for superficial area. I think it is the fairest system of measurement ever adopted by this or any other club. As showing its operations, I will give the following entries for the twenty-fifth annual regatta which was sailed June 22, 1871:

SCHOONERS.

YACHT.	OWNERS.	CUBIC FEET.
Tidal Wave	Wm. Voorhis	3269
Eva	S. J. Macy	2233
Madeleine	Jacob Voorhis, Jr.	3824
Wanderer	Louis J. Lorillard	5346
Alarm	A. C. Kingsland	5891
Columbia	Frank Osgood	4801
Idler	Thomas C. Durant	2932
Foam	Sheppard Homans	2496
Sunshine	E. Burd Grubb	850
Magic	J. Lester Wallack	2492
Dauntless	James Gordon Bennett, Jr.	7124
Tarolinta	H. A. Kent	3629
Rambler	James H. Banker	5993
Alice	George W. Kidd	1806
Sappho	Wm. P. Douglass	7431
Palmer	R. Stuyvesant	4564
Halcyon	James R. Smith	2864

SLOOPS.

YACHT.	OWNERS.	CUBIC FEET.
Breeze	A. C. Kingsland, Jr.	505
Gracie	William Krebs	1473
Ariadne	Theodore A. Strange	558
Addie	William H. Langley	1099
Vixen	Ludlow Livingston	707

It is said that after a season of as great excitement as that of 1870, the next one may be counted on as being rather dull; but this was not the case during the season of 1871. Probably the certainty that Mr. Ashbury was again coming for the cup kept the interest from flagging, and then, in addition to this, Mr. Bennett had been elected commodore, and the young element in the club was in control. It was at this regatta that Mr. Bennett's challenge cups for schooners and sloops were first sailed for, and the great controversy as to the South-west Spit buoy arose. Before this, the yachts of the club had always turned buoy No. 10, and therefore there ought not to have been a question about the matter; but as part of the fleet turned No. 10, and part No. 8½, and as if No. 10 was the right mark, the *Idler* won, while if No. 8½ was the turning-point, the *Tidal Wave* won, and as the club excursion boat went and laid at No. 8½, while the judges' tug lingered at No. 10, the matter was complicated, and was referred to Mr. Geo. W. Blunt, pilot commissioner, to the Hydrographic Office, at Washington, and to various Sandy Hook pilots.

All of the pilots said: "Although buoy 8½ is on the Spit, No. 10 is the proper Spit buoy, and if you attempt to turn 8½, with twenty-two feet of water, you will go aground." Mr. Blunt and the Hydrographic Office said, "8½ being on the Spit, is the Spit buoy," and the *Tidal Wave* was given the race, but the club since that time has ordered its yachts to turn both buoys, so that there can be no mistake.

At this regatta, there was also another innovation, the effect of the ascendency of the young and progressive element. Two prizes, one of $600 for schooners, and one of $400 for sloops, were offered open to yachts of any recognized yacht club, and for these, in addition to the sloops named above, there entered the *Peerless*, Atlantic club; the *Kaiser Wilhelm*, Brooklyn club, and the *Coming*, Eastern Club. There were no outside schooners, no other club than the New York, even down to this date, having any boats of that rig large enough to compete here.

The *Tidal Wave* took all three cups: the Bennett challenge, the subscription and the regular club cup; the *Columbia*, then brand new, being second. The sloop *Addie*, took all three prizes; the *Gracie* coming second.

This seems curious to us now, but at that time, the *Gracie* was a very different yacht from what she is at present. The judges who decided this question of the buoys were Messrs. Philip Schuyler, Stuart M. Taylor and William Butler Duncan.

On June 27, in this year—1871—the Brooklyn Yacht Club first became prominent, and at its regatta, all the principal schooners of the New York Yacht Club appeared as starters. As previously stated, Mr. Jacob Voorhis, Jr., owner of the *Madeleine*, and a millionaire, had been elected as its commodore, and had carried with him many of the New York owners. Some of them never knew of their being proposed for membership, until they received the notification of the club secretary, with a receipt for initiation fee and a year's dues, which had been paid by Commodore Voorhis.

The measurement was by the old New York rule of superficial area, under which the *Columbia* went in at 1,694 feet; *Dauntless*, 1,924, and *Sappho*, 1,979. The *Dauntless* came in first and won the prize, without allowance; but the club and union prizes were given to the *Madeleine* by three seconds. There is good reason for saying at this lapse of time, that the decision was a mistaken one, and that it was only because she was "Our Commodore's" yacht that these prizes were awarded to her, the *Columbia* having won them, beyond a doubt. The sloop *Gracie* took the prize without allowance, and the Union prize, beating the *Addie* 4m. 23s.; but not belonging to the club, she could not win the club prize, and that was captured by the *Addie*, which beat the *Kate* over 14 minutes.

The decision in favor of the *Madeleine* was the first thing which caused the decline and fall of the Brooklyn club. It was evidently so unjust that Mr. Osgood withdrew, and carried several others with him; and although the club had a seeming prosperity for a couple of years after this, it was hollow. Mr. Osgood sent the following letter to the regatta committee, Messrs. W. W. Van Dyke, Alonzo Slote, W. B. Nichols, John H. Lewis and S. P. Bunker:

GENTLEMEN: I suppose it is only necessary for me to draw your attention to the unaccountable mistake in your decision in regard to the race yesterday, to have you rectify the error. The time which elapsed between the passing of the home stake boat by the *Columbia* and the *Madeleine* is incorrectly given, being 3m. 13s., instead of 1m. 13s. The time was obtained from your own appointed time-keeper. Unquestionably, to my mind, Commodore Voorhis must be fully aware of the actual difference in the time of arrival of our respective boats, as on an occasion like this every yacht owner knows the time of his passing the home stake boat. I am prepared to furnish you with full proof to substantiate my claim of having fairly beaten the *Madeleine*.

June 28, 1871.

Two or three times, summer residents at Cape May have induced the New York yachtsmen to come down there and sail a race, and the first time that this occurred was in this season of 1871. Attracted by the offer of two $1,000 cups, one for schooners and one for sloops, open to any yacht-club in the world, several of the schooners of the New York club and sloops of that and other clubs went down. They found a miserable harbor, very difficult of entrance, and an open roadstead with poor anchorage outside; and came home vowing that nothing should tempt them there again.

These were the yachts which went down. Schooners: *Sappho, Dauntless, Rambler, Alarm, Wanderer, Columbia, Palmer, Madeleine, Tidal Wave, Madgie, Eva* and *Sunshine.* Sloops: *Gracie* and *Vindex*, of the New York, and

39⅞ miles. If the affair was remarkable for anything, it was for the sailing of the schooner *Wanderer* on the passage down. In a nice working breeze dead ahead, she beat the *Sappho* and *Dauntless* handsomely. In an all day beat, the breeze steady, she led the *Dauntless* about an hour and the *Sappho* over an hour and a quarter. And she has never sailed remarkably well since. Captain "Bob" Fish was on board of her on this occasion, and said on arrival at Cape May,

"FANNY," BOSTON.

Daphne of the Atlantic club. The race was sailed July 4, from a point off the hotels at Cape May, to and around the Five Fathom Light-ship; thence five miles northeast to a stake-boat, and back to the place of departure, a total distance of

that he could make her do better. He got permission to alter her trim, and did so, and the next day, in the race, she was nowhere.

The *Sappho* was bound up too tight. Next day, the lanyards of her rigging were eased a trifle, and she beat all the other yachts with ease.

The *Dauntless*, *Sappho* and *Wanderer* raced from Sandy Hook to the Cape May Light-ship for a $500 cup, a little private arrangement; and as stated, the *Wanderer* won. The schooner *Dreadnought* made her first appearance in this trip to Cape May, but her performance on the way down did not warrant Captain Samuels in entering her for the race, and although she started with the lot, she did not return, but bore up and ran for New York.

The *Sappho* took the Citizen's Cup with time allowance, and the Benson Cup (both $1000 mugs), beating the *Columbia* 5m. 5s. The *Gracie* took the Citizens' Cup for sloops, beating the *Vindex* 2m. 37s. The latter yacht was new, and she also embodied several new principles. She was iron, for one thing, and I think that in time, iron will supersede wood entirely for the hulls of yachts. Then, too, she had parts of the cutter rig; that is, she had the short mast and long topmast; but I think her mainsail laced to the boom, and that she had a standing jib. She had also a stay to the knight heads, and a stay-sail. Mr. Center, who designed her, afterwards had her jib to set flying, and found it a great improvement.

SCHOONER "PALMER."[1]

So far as model was concerned, the *Vindex* had little in common with the modern cutter, being over 17 feet beam on a waterline length of 56 feet. She may, however, I think, be said to have been the first American-built sloop that was cutter rigged.

During the August cruise, this year, 1871—for the first time the Eastern and New York club fleets joined each other

[1] Original owner R. F. Loper; present owner Rutherford Stuyvesant.

and, with some few exceptional years, have done so ever since. Many New York club members joined the Eastern, and some of the gentlemen of the Eastern club joined the New York. There has been a community of interest between them ever since. They joined company this year, at Newport, and sailed east around Cape Cod and had a regatta at Swampscott, Mass., for prizes, $1000 for schooners and $500 for sloops, offered by the Eastern club, and to be sailed according to its rules. There were also prizes offered by the citizens of Swampscott—$800 for schooners, and $400 for sloops, without any time allowance.

As showing that even as late as 1871, no other club than the New York was of much importance, I will give the vessels of the two clubs and their sizes. The New York club entered, schooners: *Columbia* (220); *Sappho* (274); *Dauntless* (268); *Fleetwing* (206); *Dreadnought* (275); *Idler* (133); *Wanderer* (187); *Tarolinta* (178); *Halcyon* (121); *Magic* (91); *Eva* (81); *Foam* (111); *Tidal Wave* (153); *Vesta* (201); *Sprite* (77); *Rambler* (242).

The Eastern Club had the schooners: *Rebecca* (77); *Belle* (45); *Edith* (47); *Juniata* (81); *Vivien* (52); *Ethel* (60); *Julia* (80); *Ianthe* (35); *Glimpse* (59); *Dawn* (41); *Silvie* (106); and *Zephyr* (41).

In sloops, the New York club entered the *Vixen* (32); *Sadie* (26); *Gracie* (58); and *Vindex* (61).

The Eastern club had sloops *Alarm* (21); *Alice* (24); *Coming* (54); *Violet* (15); *Narragansett* (28).

Things have changed relatively since that time, and to-day the Eastern Club has the finest club-house in America, on Marblehead Neck, and some of its schooners — the *Ambassadress, Fortuna, Gitana*, etc. — are the peers of any in the world; while the *Puritan, Thetis*, and a half-dozen other big sloops cannot be beaten by single-stick vessels anywhere.

This course at Swampscott was 39¼ miles in length, and there started thirty-three yachts, of which thirty finished the course. It was the largest number which had ever competed in American waters.

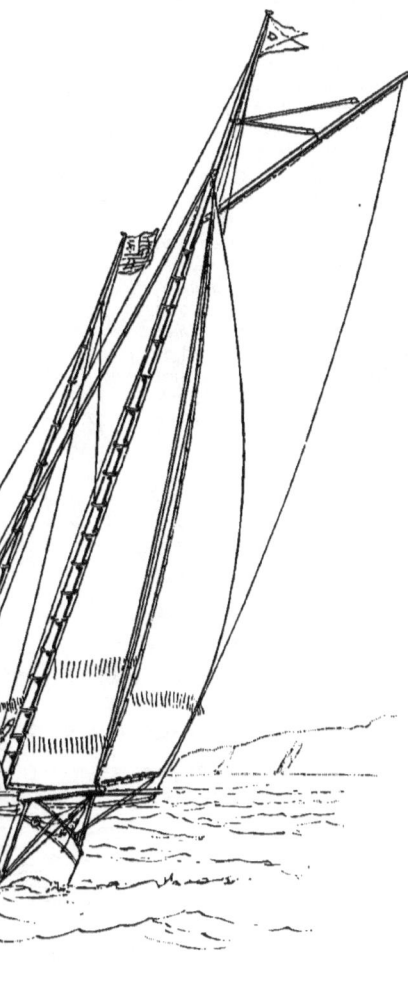

"FROLIC," SAN FRANCISCO.

Now-a-days we think nothing of starting over a hundred. The *Columbia* took the Eastern Club prize, and also the Swampscott, and the *Gracie* took both of the sloop prizes. The wind was moderate from east-south-east.

The cup awarded to the *Tidal Wave* on the occasion of the muddle about the buoys 8½ and 10, was not retained by the owner of that schooner. He returned it to the club, and it was again raced for over the Block Island course, August 21, 1871; the conditions of the deed of gift providing that it may be competed for over either of these club courses; and to make the matter interesting, the flag officers subscribed for a cup for sloops. Eight schooners and four sloops started, and the prizes were won by the schooner *Madgie* and the sloop *Sadie*. The *Sappho* made the best time, but was beaten 45½s. by the *Madgie* on time allowance.

August 22, 1871, an attempt was made to sail for the Douglass $1,000 Cup over the 64-mile course off Newport, and there started the schooners *Wanderer*, *Alarm*, *Dauntless*, *Dreadnought*, *Palmer*, *Tidal Wave* and *Madgie*. Only the *Palmer* and *Dauntless* were timed at the finish, and they did not arrive until after nine o'clock in the evening, long after the nine-hour limit had expired. August 24, a start was made for the Lorillard Cup of $1,000 over the long course, but the chance of doing the distance in nine hours was so remote, that only the *Sappho*, *Palmer* and *Dreadnought* put in an appearance at the starting line, and a thick fog prevailing, the judges decided not to start the boats. There was, however, a good breeze, and Vice-commodore Douglass determined to try the *Sappho* over the course alone. I was fortunate in

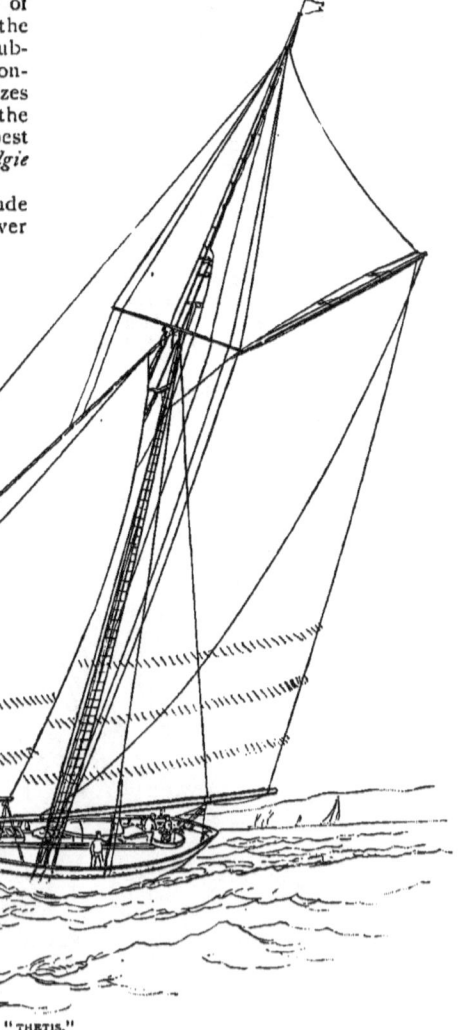

"THETIS."

having been on board of her on this occasion, and enjoyed a most beautiful sail, the yacht having made the course in less time than it had ever, as matter of record, been done before. She started at 12h. 12m., with wind south-west, and we beat down to the buoy, heading south on one tack and west on the other, the yacht going a nice clean full, and doing better, I thought, than if she had been racing. That celebrated racing skipper, "Sam" Greenwood was at the wheel, and he had a way, I thought, of pinching the *Sappho* too much. She would work in seven points; but for her best work she required eight. Given a good clean full within four points of the wind, she was the smartest vessel in the whole world. On this occasion she rounded the Block Island buoy at 3h., 19m., 10s., and, running with only working topsails, balloon jib topsail and main topmast stay-sail for light kites, she turned the Sow and Pigs Light-ship at 6h., 30m. She could then just lie her course for the finish, and arrived there at 8h., 0m., 30s., having made the 64 miles in 7h., 48m., 30s., beating the record.

There was another trial for the cup over this course August 25, the *Sappho, Dreadnought* and *Madgie* starting. The *Madgie* withdrew when a part of the course had been covered, and the other two made the course, but not in nine hours. Their times are worth giving, as showing how close the *Dreadnought*, on this occasion, came to the *Sappho*. They were, *Sappho*, 10h., 12m., 0s.; *Dreadnought*, 10h., 18m., 4s.; a difference of 6m., 4s., in favor of the *Sappho* on elapsed time, and in a race of that distance on allowance, she would have won.

I have said that the Brooklyn Yacht Club was about this time coming into some prominence; but without the aid of the New York it did not make much of a show, even as late as August, 1871. It had its cruise at the same time as the New York club had, and it had a regatta at New London, August 25, 1871. The only schooners it could boast were the *Madeleine* (the flag-ship) and *Fleur de Lis*, at that time owned by Mr. John R. Dickerson, and the present owner of the *Madeleine*; and of its sloops were the *Addie, Qui Vive, Kate, Kaiser Wilhelm, Maggie B., Sophia, Mary, Recreation, Jennie, Twilight, West Wind, Nettie B., Ada, Nettle, Frolic, Twilight, Carrie, Khedive, Water Lily, Ah Sin, Haidee,* and *Annie.* Many of these were yachts belonging to New London. The *Maggie B.,* at this time, was owned by the celebrated "Tom Thumb," who was a member of the Brooklyn club, and had a

"INTREPID."

racing crew of Bridgeport fishermen, who made the *Maggie* very hard to beat. Their diminutive owner was immensely popular with them.

The *Sophia*, another of these sloops, had a most melancholy ending, having capsized and sunk in the Sound, a few years ago, with loss of several lives.

I mention this regatta of the Brooklyn club, because my readers may have thought the title of these chapters a misnomer, and that instead of being a history of American yachting, it was simply a history of the New York Yacht Club. But in point of fact, down to this time, there was little else of American yachting save the New York club, although the Eastern and the Brooklyn, fostered and encouraged by the New York, were coming into prominence.

It was on this cruise of the New York club, and while at Newport, that it received Mr. Ashbury's proposition to come here in the schooner *Livonia* for the *America's Cup*.

As that gentleman has been somewhat misrepresented, I will state exactly what his proposition was. He was to come representing twelve different clubs, and in his letter he says distinctly: "If the *Livonia* shall win *a majority* of the races, the cup would then go to the club under whose flag I sailed in the last and final

"HOPE."

race;" and he wanted a series of twelve races. He has been represented as desiring to sail twelve races, and if he won *one out of the twelve*, to take the cup. I feel like saying that Mr. Ashbury was not treated over and above fairly by the New York club, and am glad to have him set right on this important point.

As I have said, the club received this proposition at a regular club meeting held on the *Dauntless*, and some of the members were in favor of replying and rejecting a proposition which no one at the meeting ever dreamed of accepting. Others, however, said, "Let him come and we'll make terms with him after he gets here," and Mr. Ashbury was induced to bring the *Livonia* here under the impression that the propositions contained in his letter of August 12, which was submitted at this meeting, had been accepted.

The last race of this brilliant series at Newport, was for the usual cup presented annually by the citizens and valued at $1,000, which was sailed August 28, over the usual Block Island course. There were nine starters, and with the usual moderate south-west wind, the *Palmer*, at the Block Island buoy, had a long lead and looked a sure winner; but in gybing around the buoy, one of Mr. Stuyvesant's guests was taken overboard by the main sheet. He swam towards the stake boat anchored near the buoy, and shouted to Mr. Stuyvesant to go on, but that gentleman refused to do so, and rounded to and took him on board again, thus giving away her chance for this splendid prize, which was finally won by the *Sappho*, beating the *Columbia* 3m. 8s.

October 2, 1871, after the return of the club to New York, the sloop *Gracie* challenged the *Addie* for the Bennett Cup over the New York course, and won it by 22s. Then the yachts or some of them went on to Newport again to sail the unfinished races for the Douglass and Lorillard Cups over the long 64-mile course. That for the Lorillard Cup was sailed October 9, 1871, the starters being the *Enchantress* — then owned by Mr. George Lorillard — the *Palmer*, *Dreadnought*, and *Sappho*. The *Dreadnought* carried away flying jib boom before the start and ran back to the harbor. The *Enchantress* struck a sunken rock or wreck running from the Block Island buoy to the Sow and Pigs. The other two kept on, and the *Sappho* won, making the races in 7h. 24m. 58s., and beating her own record. There was a moderate gale from south-west.

October 10, the unfinished race for the Douglass $1,000 cup was sailed over the Newport long course, in a fresh northerly breeze; the starters being the *Dreadnought*, *Palmer*, *Madgie*, and *Wanderer*. It was in this race, that the *Dreadnought* immortalized herself, beating the *Palmer* and finishing the race in 7h. 33m. 58s., coming within nine minutes of the *Sappho's* time in the preceding race. This concluded the racing for this year, except the *Livonia* races for the *America's* Cup and those which grew out of the visit of that yacht to this country, and these I will reserve for a future chapter.

THE HISTORY OF AMERICAN YACHTING.

BY CAPTAIN R. F. COFFIN,

Author of "Old Sailor Yarns," "The America's Cup," etc., etc.

IV.

FROM 1871 TO 1876.

WHEN Mr. Ashbury came here with the schooner *Livonia*, in 1871, for the *America's* Cup, I don't think that he had much hope of winning it, for the performance of his yacht in the British waters had not developed any speed superior to that of the *Cambria*, and that schooner — as had been abundantly shown — was inferior to most of the American yachts; but Mr. Ashbury had built the *Livonia* expressly for this service, had challenged for the cup, and his challenge had been accepted, and with true British perseverance, he determined to see the thing through. Then, too, it was said, that he was desirous of being returned to Parliament, and naturally expected that a recognition of his pluck and enterprise would materially assist his canvass, whether he was successful in his quest for the cup or not. In this he was probably correct, for on his return after these races he became M. P. for Harwich.

The *Livonia* arrived here October 1, 1871, after a passage of nearly twenty-nine days, and after considerable correspondence, the match was finally made to consist of the best four out of seven races, three of which were to be over the club course and four over a course twenty miles to windward (or leeward) from the Sandy Hook Light-ship and return. It is not necessary to go into the details of these races. The club committee selected the keel schooners *Sappho* and *Dauntless*, and the center-board schooners *Columbia* and *Palmer*, reserving the right to name either of these four as a competitor for the *Livonia* on the morning of each race. I will give the size and ownership of the five yachts.

NAME.	OWNER.	DISPLACEMENT.	APPORTIONMENT.
Livonia.....	James Ashbury.........	6,651	1,881
Dauntless...	James G. Bennett, Jr...	7,124	1,924
Sappho.....	W. P. Douglass.........	7,431	1,951
Palmer.....	Rutherford Stuyvesant..	4,546	1,659
Columbia...	Franklin Osgood........	4,861	1,694

Mr. Ashbury had vainly protested against the selection of four vessels, claiming that as he had but one, so only one should be put against her for the whole series of races; but he finally yielded this point, and the races were sailed, three between the *Columbia* and *Livonia* and two between the *Sappho* and *Livonia*. The dates were October 16, 18 and 19 with the *Columbia*, and October 21 and 23 with the *Sappho*. The only race won by the *Livonia* was the third, on October 19, over the club course, in a fresh breeze from west-south-west. The *Columbia* ought not to have been started in this race, as her crew were worn out by the race of the previous day, and her owner and officers had not supposed that she would be again selected. She carried away her flying jib stay, when rounding the Spit buoy, going out, and this causing her to gripe badly, her steering gear gave out on the return. She was beaten 19m. 33s. actual time, and 15m. 10s. corrected time.

There was a dispute as to the second race, which was over the outside course, the captain of the British yacht, believing that he had to leave the outer mark on the starboard hand, gybed his boat around it in order to do so. The captain of the *Columbia*, who had, previous to the start, been informed that he could turn it either way, luffed around and secured an advantage by doing so. Mr. Ashbury asked that this race might be thrown out and another sailed in place of it, but the committee refused. I was on board of the *Columbia* during this race, and I think that Mr. Ashbury's request was a proper one. The *Livonia* had led the *Columbia* all the way to the outer mark, and but for this misunderstanding, would have begun the return in the lead. It was a straight

"ISIS."

reach to the finish, and it was possible for her to have won. As the committee was clearly at fault in not giving explicit directions as to how the mark should be turned, and as they had given permission to one skipper to turn either way, and had not given the same permission to the other, and as it was evident that the captain of the *Livonia* — following the racing rule of England, which provides that all marks shall be left on the starboard hand unless other direction be given — had lost time, Mr. Ashbury's request was a reasonable one, and should have been granted. The committee were Moses H. Grinnell (chairman), Sheppard Gandy, Robert S. Hone, Philip Schuyler and Charles A. Minton. In support of its decision, the committee called attention to the protest made by the owner of the yacht *Brilliant* against the *America* in the original race for the cup, claiming that the *America* had gone the wrong side of the Nab Light vessel, and the British committee decided that the *America*, having no written instructions, could go either side. But this was hardly a parallel case, and in this *Columbia-Livonia* matter the committee were so clearly at fault in omitting the written instruction that Mr. Ashbury's very reasonable request for another race should, I think, have been granted. It should be stated, however, that Mr. Ashbury's first claim was that the race be awarded to the *Livonia*, and this the committee was right in refusing.

After the final race with the *Sappho*, however, October 23, Mr. Ashbury sent a communication to the committee, informing them that the *Livonia* would be at her station the next day for the sixth race, and also on the following day for the seventh, he claiming that he had already won two races, and that these two would give him the cup. The committee sent no answer to this, and the *Livonia* went over the course in a race with the *Dauntless* for a fifty-guinea cup, the match being sailed under the old 1860 measurement, the yachts being entered as follows: *Dauntless*, 2,899 square feet; *Livonia*, 2,512 square feet. They went to windward from the light-ship, and the *Dauntless* won by 11m. 3s. actual and 6m. 3s. corrected time. By the new system she would win by 10m. 31s.

Mr. Ashbury did not carry out his promise of sailing over alone on the next day, and this match ended his racing in America. Either one of the four schooners selected could have beaten the *Livonia* always in any square race.

There were two more ocean races as a wind-up to the season of 1871: the *Sappho* and *Dauntless* each sailing a match with the *Dreadnought* for $250 cups, and each beating her with ease. The *Dreadnought* was built originally for Mr. Frederick W. Lane, who, I believe, never went on board of her. Before she was finished, he had altered his mind, and concluded that he did not want a yacht, and captain Samuels, under whose superintendence she was built, was running her during this season in order to find a customer for her. She was afterwards purchased by Mr. A. B. Stockwell, and by him sold to the late C. J. Osborn, who had Mr. Henry Steers lengthen her, making her a much faster yacht than before. I may say, in concluding the events of the year, that Mr. Ashbury left for England, October 30, on the Cunard steamer, and Captain Wood took the *Livonia* home. At the first meeting of the New York Yacht Club in the year 1872, a letter was read from Mr. Ashbury, in which

"ESTELLE."

he charged the club with unsportsmanlike conduct, and the club at once ordered that the cups he had left to be sailed for should be returned to him, and that ended his connection with American yachting.

From this time on, yachting events have been too numerous to mention them in detail, and I shall only refer to the most important. It was in this year that the Atlantic club began to come into prominence, with Mr. William Voorhis as its commodore, and the schooner *Tidal Wave* as its flag-ship; and this year at its annual regatta it started three such schooners as the *Tidal Wave*, *Resolute* and *Peerless*, with ten sloops, those in the first class being the *Gracie*, *Addie*, *Orion* and *Vixen*.

The regatta of the New York Yacht Club, this year, on June 20, was remarkable from the fact that the flying start was adopted — the fairest way of starting yachts which has yet been tried — and it was also remarkable from the fact that in an exceptionally fine lot of schooners, the little *Ianthe*, the very slowest of the lot, beat all of them without allowance of time. When all except her had been out around the light-ship and were returning, they met her at the bar buoy going out. They all got becalmed in the bay, and with a strong flood drifted away to the westward, while the *Ianthe*, with a cracking breeze, went out to the light-ship and returned, and keeping in a little cat's-paw of wind, luffed over close to Coney Island Point, and went on up to the finish-line, distancing the lot. The *Peerless* took the schooner prize on allowance of time, and the winning sloops were the *Gracie* and *Vixen*, the prizes being four $250 cups.

It was about this time that Mr. Lester Wallack, the actor, began to come into prominence as a patron of yachting, and he gave a cup for schooners, which was competed for June 24, 1872, the course being from off No. 5, at the point of Sandy Hook, to a stake boat close in to Long Branch, where Mr. Wallack had a cottage. There was a good entry, and the *Madeleine*

"CLYTIE."

"GRAYLING."

secured the prize, beating the *Peerless* 7m. 10s. It was in this year, also, that the sloop *Meta* was built by Mr. P. McGiehan, at Pamrapo, N. J., for Mr. G. H. Beling, and with Captain Ellsworth and his Bayonne crew on board, she took rank as the fastest single-stick vessel in the country. Mr. Beling, during the previous year, had the sloop *Kaiser Wilhelm* built by Mr. McGiehan, and there was enough of suspicion that during some of her races ballast had been shifted, to cause her owner to be black-balled in the New York club, and on this account he adopted as his signal on the *Meta* two black balls, and entered her for all the races that were possible. July 23, 1872, the *Meta* sailed a match with the *Gracie* from buoy No. 5, twenty miles to windward and return. The *Gracie* at that time was owned by Mr. S. J. Colgate, New York Yacht Club, and was 58 feet 5 inches long. Last year she was 79 feet 10 inches, and this season is still longer. The *Meta* was 61 feet. The race was sailed under the rules of the Brooklyn club, and the *Meta* beat the *Gracie* in actual time 31s., but on corrected time the *Gracie* won the race by 1m. 45s.

I may mention in passing that beside the Brooklyn and Atlantic clubs, there had now come into prominence the Harlem club, of which one of the ruling spirits was Mr. Harry Genet, the brilliant politician of the Tweed régime. "Prince Hal" he was called, and the house on the point at Port Morris, now occupied by the Knickerbocker club, was built under his reign, which I may say was short and brilliant. But in this year, 1872, the Seawanhaka club was a year old. It had its annual regattas at Oyster Bay on each recurring Fourth of July, and they were the most enjoyable yachting events of the year.

The Jersey City club, too, began to loom up prominently. Their regattas were social affairs, and used to take place at Greenville, N. J., from a tavern called the "Idle Hour." After the race of the yachts, a banquet was served, ladies were present in great number, and the affair wound up with a dance in the evening. Apropos of

"FORTUNE," BOSTON.

Mr. Ashbury, the Havre (France) regatta was sailed July 12, 1872, and among the entries were the British schooners *Guinevere* and *Livonia* and the American schooner *Sappho*. The *Guinevere* was withdrawn, and Mr. Douglass, owner of the *Sappho*, at once withdrew her, declining to sail against Mr. Ashbury. He started, however, fifteen minutes after the *Livonia;* came up with and ran through her lee, and then went on over the course, finishing an hour and a half ahead of her.

In this year, 1872, it was, that Commodore Bennett presented his Brenton's Reef Challenge Cup, valued at $1,000, for an international trophy. This is the cup that was won last season by the British cutter *Genesta*. As there seemed some disinclination to entering for it, Mr. Bennett offered an extra prize of $500 in case five yachts started, to be presented to, and held by, the winner as his own private property. The only starters, however, were the *Rambler*, then owned by Mr. J. Malcolm Forbes, who now owns the sloop *Puritan* and the *Madeleine*, then owned by Commodore Jacob Voorhis, Jr., of the Brooklyn club. The start was made July 25, 1872, and the yachts had dirty weather. The *Madeleine* had to put in to New London, and did not reach the outer mark. The *Rambler* made the course in 39h. 55m. 59s. She belonged to the Eastern club, which from that time on, has constantly increased in importance.

During the August cruise of the New York club, this year, there was a handicap race over the regular Block Island course; which is important, as showing how the yachts of that time were rated. I think the clubs of both countries will finally come to this as the fairest way of sailing yacht races. Certain it is, no system of measurement has ever been satisfactory, and probably none ever will be. Here is the way in which the committee rated the yachts of that day. The schooner *Columbia* allows the *Madeleine* 1m.; *Resolute* 1m.; *Tidal*

Wave, 2m.; *Viking*, 3m.; *Madgie*, 4m.; *Magic*, 4m. 30s.; *Foam*, 5m.; *Ira*, 8m.; *Alice*, 9m. And at this rating the *Columbia* won, showing that the committee had estimated her correctly. The *Foam* was second and *Madeleine* third. The *Foam* was ruled out for fouling a stake boat, and the *Madeleine* took second prize.

The race for the City of Newport Cup, this year, was sailed over the course from Brenton's Reef to the Sow and Pigs Light-ship and return, and was the first race ever sailed over this course. It was a memorable one, being sailed in a thick fog. The schooner *Dauntless* ran into the Sow and Pigs Light vessel and carried away her lanterns. On the return, the *Madeleine* overran her reckoning, and narrowly escaped wreck on Beaver Tail. The *Tidal Wave* got close in to the beach, just east of Brenton's Reef, and had to let go both anchors, her stern just clearing the breakers, as she swung to the chains. The *Magic* collided with another schooner, and was much damaged, and as a result of all this, no schooner finished. The sloop *Meta*, of the Brooklyn club, with Capt. "Joe" Ellsworth at the wheel, made the race and won the sloop prize. There were other interesting races, but I may not stop to mention them all. The *Madeleine* challenged the *Rambler* for the Brenton's Reef Cup, and another start for this was made September 19, the course being from the Brenton's Reef Light-ship to the Sandy Hook Light-ship and return. The *Rambler* again won, her time being 43h. 25m. 32s. against 47h. 18m. 41s. for the *Madeleine*. They had heavy weather, and the *Madeleine* was much damaged in rigging.

It will be remembered that the Bennett Challenge Cup for sloops, over the regular course, was first won by the *Addie* in June, 1871, and captured from her, in October, by the *Gracie*. Among the sloops built in 1872 was the *Vision*. She was built by Mr. J. McGarrick, at the foot of Thirty-fourth street, Brooklyn. She was probably the shoalest craft ever built; a skimming dish of the skimming dishes. Her dimensions were: over all, 66 feet; water-line, 52 feet 4 inches; beam, 20 feet 9 inches; depth, 5 feet 11 inches; draught, 5 feet 9 inches. But in that day the skimming dish was the favorite model, and the *Vision* enjoyed a reputation of "fastest in the fleet." She was, in fact, enormously fast in smooth water, her great beam enabling her to carry a powerful spread of canvas. In a sea way, however, she was good for nothing. She, however, challenged the *Gracie* for the Bennett Cup, and the race came off September 20, 1872, in a howling gale from west-north-west. Had the judges sent them over the regular course, I presume both would have had to be towed in from the Light-ship; but an easier course was agreed upon, and from the Narrows they went down around the Spit buoy; thence back to Craven Shoal buoy, returning over the same course and finishing at Craven Shoal. After rounding the Spit buoy the second time, in trying to get the main sheet aft, five of the *Gracie's* men were taken overboard, and she had to stop and pick them up, after which she anchored for the night in the Horseshoe, and the *Vision* won in 4h. 25m. 55s. Both yachts sailed with three reefs tied down.

On October 10, 1872, there was a race for the Bennett Cape May Challenge Cup, the one now held by the *Genesta*, the starters being the *Dreadnought* and *Palmer*, and the *Dreadnought* won. The times were *Dreadnought*, 25h. 05m. 40s.; *Palmer*, 26h. 45m. 5s. I may mention in passing, as it is a species of yachting, that it was in 1872 that the first canoe club was formed, and that it sailed its first regatta October 19, in Flushing Bay, and also, I may state, that it was in this year that miniature yachting was inaugurated, and for two or three seasons flourished very successfully at the Prospect Park Lake, in Brooklyn. It is to be regretted that it did not continue popular, as I think some of the improvements in model and rig of the larger pleasure craft may be directly traceable to the experiments with the four and five feet models on the park lake.

It will be interesting also to state, as showing how the sport was being developed in this country, that at the beginning of the season of 1873, the Boston club had enrolled thirteen schooners, twenty-five sloops and two steamers. The Eastern club had thirty-two schooners and thirteen sloops. The South Boston club had four schooners and twenty-seven sloops. The Dorchester club had six schooners and fifty-one sloops. The Lynn club had one schooner, twenty-three sloops and eight cat-rigged boats. The Beverly club had thirty-seven sloops and cat rigs, and the Bunker Hill club had five schooners, thirteen sloops and one steamer. In this neighborhood, beside the New York, Brooklyn, Atlantic, Seawanhaka and Jersey City, there were the Williamsburgh, the Harlem, the Long Island, the Bayonne, the Columbia, the Pavonia clubs, and four or five other minor organizations, so that, as will be seen, the sport had broadened out immensely.

Under date of April 18, 1873, the Secretary of the Royal Western Yacht Club of Ireland sent a letter to the New York Yacht Club, informing it that Her Majesty, having graciously given the club a cup to be sailed for at the annual regatta on July 30, the club would be pleased if the members of the New York club or any of them would compete for the same. So far as I can recollect, however, no American yacht accepted this courteous invitation.

No one who was fortunate enough to witness it, will ever forget the glorious finish of the annual regatta of the New York Yacht Club in 1873, when the *Madeleine* went over the line a winner just as a hard squall from the north-west struck the fleet, the *Madgie* being very nearly capsized. The *Madeleine's* time was the best ever made over this course, the start and finish being in the Narrows. Her actual time was 4h. 1m. 20s., and her corrected time was 3h. 57m. 43s.

The sloops *Meta* and *Vision* sailed a match outside of Sandy Hook twenty miles to leeward from buoy No. 5 and return, which was the first Sunday race ever sailed by a yacht of the New York Yacht Club. The race was for $500 a side, and the *Vision* won by 7m. 32s. The *Meta* lost her topmast, but as the other boat had to house hers on the return, this did not injure her chance for the race in any way.

The cruise of the Brooklyn club, this year, was the most memorable in its history, and from this time it steadily declined in importance. The cruise of the New York Yacht Club, also this year, will long be remembered, the fleet coming out of Glen Cove harbor in the early morning, with a light easterly wind, which gradually increased to a reefing breeze from north-east, and scattered the fleet, forcing the yachts to seek harbors wherever possible. The only ones which got through to New London, where they were bound, were the schooners *Idler* and *Rambler*, and the iron sloop *Vindex*. It was four days before the fleet was united, a hard gale from north-east prevailing most of the time.

One of the most memorable races ever sailed in this neighborhood was that between the open sloops *William T. Lee* and *Brooklyn*, from the head of Gowanus Bay to and around buoy No. 8½ and return, for a wager of $1,000. The yachts were about 28 feet in length, and I presume the sail spread from the luff of the jib to the leech of the mainsail was full 65 feet. The wind was quite fresh from south-west, yet both boats managed to carry whole sail for the entire race, and the wonder of it was that neither capsized. Both yachts, however, were terribly strained, and leaked like baskets at the conclusion of the contest. Such races as that we are not likely to see again. The occupation of the professional sailing master is gone probably forever. Owners of boats like the *Lee* and *Brooklyn* have banded together in clubs which have, as a rule, adopted the fixed ballast regulation, and such rigs as were carried on the old time open racing sloop, are out of the question. It was in October of this year, that Commodore Bennett offered his celebrated prizes for yachts, pilot boats, working schooners and fishing smacks, to be sailed for from Owls' Head to and around the Five Fathom Light-ship off Cape May. $1,000 was offered to the winning yacht vessels of any club being eligible to entry; $1,000 was offered to the winning pilot boat; $250 to the winning working schooner, and $250 to the winning smack. The starters were schooner yachts *Enchantress*, *Alarm*, *Clio*, *Eva* and *Dreadnought*. The pilot boats were: the *Widgeon*, *Hope*, *James W. Elwell*, *Thomas S. Negus*, *Edmund Blunt*, *Mary E. Fish*, and *Charles H. Marshall*. Working schooners, *W. H. Van Name* and *Reindeer*, and schooner smack *Wallace Blackford*.

The vessels had pretty heavy weather on their return from Cape May. The yacht *Enchantress* was the first to return, followed two hours later by the pilot boat *Thos. S. Negus*. The *Van Name* got the working schooner prize, and the smack, *Wallace Blackford*, the $250 for a sail over alone.

Commodore Bennett had, the previous year, presented to the club, five valuable cups, and from first to last, during his connection with the New York club, he has contributed to it probably in prizes as much as all its other special contributors put together. I don't mean the total of the regular club prizes paid for out of its revenues, but the special contributions, such as the Douglass or Lorillard Cups, etc.

Rather an amusing episode in connection with Mr. Bennett's Cape May Cup occurred during the season of 1873. As I have stated, it was first raced for by the *Palmer* and *Dreadnought*, and won by the latter. Mr. J. F. Loubat, the owner of the schooner *Enchantress*, challenged for it, the date of the race being fixed for October 14, but the *Dreadnought* was so much damaged on October 11, in returning from the Cape May race alluded to above, that her owner, Mr. A. B. Stockwell, asked for an extension of time to enable him to repair damages. Under ordinary circumstances, this would have been granted, but on October 15, the season, so far as this Cape May Cup was concerned, ended, and Mr. Lowbat, thinking that the story of damage to the *Dreadnought* was but a ruse to get the race off for the season, refused to grant the delay, and appeared on the day appointed and sailed over the course, taking the Cup, and also a $1000 check from Mr. Stockwell, the amount of the little private wager between the two gentlemen. There was an amusing newspaper controversy, and the matter ended by Mr. Lowbat sending the check to a charitable institution, and returning the cup to the club later on.

I see nothing of special note during the year 1874, until October 13, when the great match race between the *Comet* and *Magic* took place, over the New York Yacht Club course. The race was ostensibly for the possession of the Bennett Challenge Cup, held by the *Comet* and challenged by the *Magic*; but there were many outside bets, and probably as much as $100,000 changed hands on the result of this race. The *Magic*, at that time, was owned by the late Mr. Wm. T. Garner, who afterwards lost his life in the heroic endeavor to save that of his wife, when the ill-fated *Mohawk* capsized off Staten Island, and it was reported that he won sufficient on this *Magic-Comet* race to pay for the hull and spars of the *Mohawk*, and that but for the victory of the *Magic* on this occasion, the *Mohawk* would not have been contracted for.

The owner of the *Comet*, Mr. Wm. H. Langley, the late Jacob Voorhis, Jr., Commodore of the Brooklyn club, and many of the members of that club who had confidence in the *Comet*, with Capt. Joe. Ellsworth at the wheel, lost very heavily on this occasion. There is slight doubt but that the *Comet* was the smarter of the two schooners, but Mr. Langley underrated his

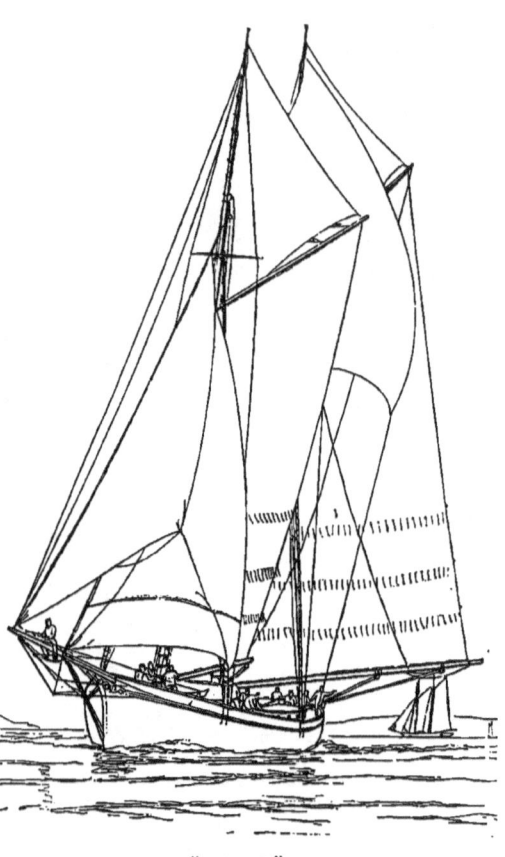

"ADRIENNE."

adversary, and neglected to put his yacht in as perfect form as she might have been. On the other hand, Mr. Garner gave his captain *carte blanche* as to expense, and the *Magic* started with a splendid lug-foresail, and a crew of twenty-five men. I was fortunate in having been a guest of Mr. Garner during this race, and am certain that it was this perfect preparation, rather than any superior speed, which gained the victory for the *Magic*. The *Comet* led to the light-ship, but upon a wind, her sails did not sit as well as those of the *Magic*, and she had to give place to her. It was, however, a very close race throughout, and up to the time that the *Magic* passed through the Narrows, on the return, it was "any body's race." She managed, however, to just squeeze by Fort Lafayette on the last of an ebb tide, and reached the slack water on the Long Island shore, while the *Comet* was reaching in the strong ebb of the Narrows. I will give the dimensions of the two yachts:

NAME.	OWNER.	W. L. LENGTH. FT.	BEAM. FT.	CUBICAL CONTENTS. FT.
Magic..	Wm. T. Garner...	78.85	20.9	5,077.79
Comet..	Wm. H. Langley .	73.03	21.95	4,662.44

The *Magic* allowed 2m. 14s.

was a miserably rainy afternoon, the yacht stuck in launching, and the elegant assemblage of ladies and gentlemen whom Mr. Garner had bidden to the launch, and for whose accommodation he had chartered a large steamboat, had a moist and disagreeable time. To give an idea of the enormous sail-spread of this famous schooner, I may state that from the top of her club top-sail sprit to the water was 163 feet, and from the end of her main boom to the end of her flying jib boom was 235 feet; and withal, she was the stiffest yacht I was ever on board of. She was 121 feet on waterline, 30 feet, 4 inches in beam, and 9 feet, 4 inches depth of hold. I have sailed in her, carrying three whole lower sails, with the water just bubbling along the lee-plank-sheer, when all other yachts in company were double-reefed and staggering along with lee-rails under. The *Mohawk* had greater initial stability than any yacht ever built in this country, and only the grossest stupidity caused her to capsize. In the minds of those ignorant of the principles of nautical construction, her mishap created a prejudice against center-board vessels, entirely unreasonable, and it has not entirely been dissipated to this day. People forget that the center - board schooner *Vesta*, the center-board

"VANITA."

June 9, 1875, the unfortunate schooner yacht *Mohawk* was launched from the yard of her builder, Mr. Joseph Vandeusen, foot of North Seventh street, Brooklyn, E. D. It sloop *Silvia* and others have gone safely across the Atlantic and returned—that two-thirds of all the American coasting fleet are center-boards, and make their passages safely along the coast at all seasons of the year; and blinded by an ignorant prejudice, cry out that the center-board vessel is a death-trap, and that the slow and clumsily-working keel boat must be adopted, because she is safe.

It is interesting to note that June 14, 1875, was the date of the first Corinthian race of the Seawanhaka club in this city, and its course was around the light-ship, the start and finish being off Stapleton, S. I. The only starters were the sloops *Addie*, *Vision* and *Coming*. The *Addie* won the race.

This club had also, on June 24, an ocean match, in which yachts from the New York, Atlantic, Brooklyn and Eastern clubs were invited to enter, and it was notable as being the first match of this kind ever sailed. On July 1, in this year, there was a race around Long Island between the steam yachts *Ideal*, owned by Mr. Havermeyer, and the *Lookout*, owned by Mr. Jacob Lorillard. It was not the first race of steam yachts, for at the annual regatta of the New York Yacht club this year, a prize had been offered for steamers, and there started the two above mentioned, and the *Lurline*, the latter, winning. On this race around Long Island, however, the *Ideal* won with all ease, her time being 18h. 22m. 45s.

The Seawanhaka club had not finally deserted its old quarters at Oyster Bay, and as had been its custom heretofore, it had its annual regatta there on the 4th of July.

The yachts of the New York club were this summer again tempted to visit Cape May and sail a race there. The only thing remarkable about it was, that it was the first race of the schooner *Mohawk*, and she was beaten by the *Madeleine*, *Idler*, *Rambler* and *Resolute*.

The three large clubs, as usual, had their annual cruises over the old course, through Long Island Sound and as far east as Vineyard Haven, but there was nothing of much note occurred. After the return, however, on September 15, the *Madeleine* and *Mohawk* sailed a match over the New York club course, and the *Madeleine* won by over 9m., and on September 21, Mr. Garner, the owner of the *Mohawk* published a challenge, offering to sail any yacht, keel or center-board, twenty miles to windward and return, outside of Sandy Hook, which at once met with response from Mr. Bennett, who offered to sail the *Dauntless* against the *Mohawk* twenty miles outside of the light-ship and return for $1,000 a side, or from Brenton's Reef to the light-ship, for $5,000 or $10,000.

This correspondence brought out other challenges. At this time, Mr. Rufus Hatch had the schooner *Resolute* under a charter from her owner, and he published a challenge, offering to sail the *Resolute* against any schooner, yacht, keel or center-board, any day in October when there is an eight-knot breeze either over the regular club course, or from Sandy Hook Light-ship to Cape May Light-ship and return. Mr. Bennett accepted a race for the *Dauntless* over the long course. Mr. J. D. Smith named the *Estelle* for a race over the club course, and Mr. W. H. Langley named the *Comet* for a race over the same course, while Mr. J. M. Mills named the *Vesta*, and Mr. C. J. Osborne named the *Dreadnought* for races over the Cape May course. The stakes sailed for were with the *Comet*, a $500 cup; with the *Estelle*, *Vesta* and *Dreadnought*, dinners of twenty covers, and with the *Dauntless*, a race for the honor of the contest.

And all these challenges were the result of an editorial published in the *New York Times* of September 21, 1875, the writer of which knew probably less about yachts or yachting than he did about Sanscrit. In proof of this, I am tempted to quote from it. He says that "The center-board is an admirable device when applied to small sail boats intended for shallow waters, no one denies. When, however, a foreign-built yacht comes here and sails half a dozen races with crack center-board yachts, and we find as a result, that while the foreigner has not started a rope or sprung a spar, her competitors are so strained as to be no longer seaworthy without undergoing extensive repairs; it is pretty clear that a fleet of center-board yachts will not gain much reputation outside of the quiet waters of New York Bay or the Sound. Of course, every addition of a new center-board yacht to a fleet increases the influence of the advocates of that style of vessel, and is, hence, to be regretted by yachtsmen who prefer salt water to fresh. It is possible, however, that the leeway made by the *Mohawk* in her recent race with the *Madeleine*, in spite of her enormous center-board, will have ultimately its effect in inducing yachtsmen to doubt whether a flat-bottom and a center-board are precisely the sort of thing to be desired in a schooner of 200 tons and upwards."

Of course, every one who had any knowledge on the subject, knew that the damage to which this gentleman alluded, as having

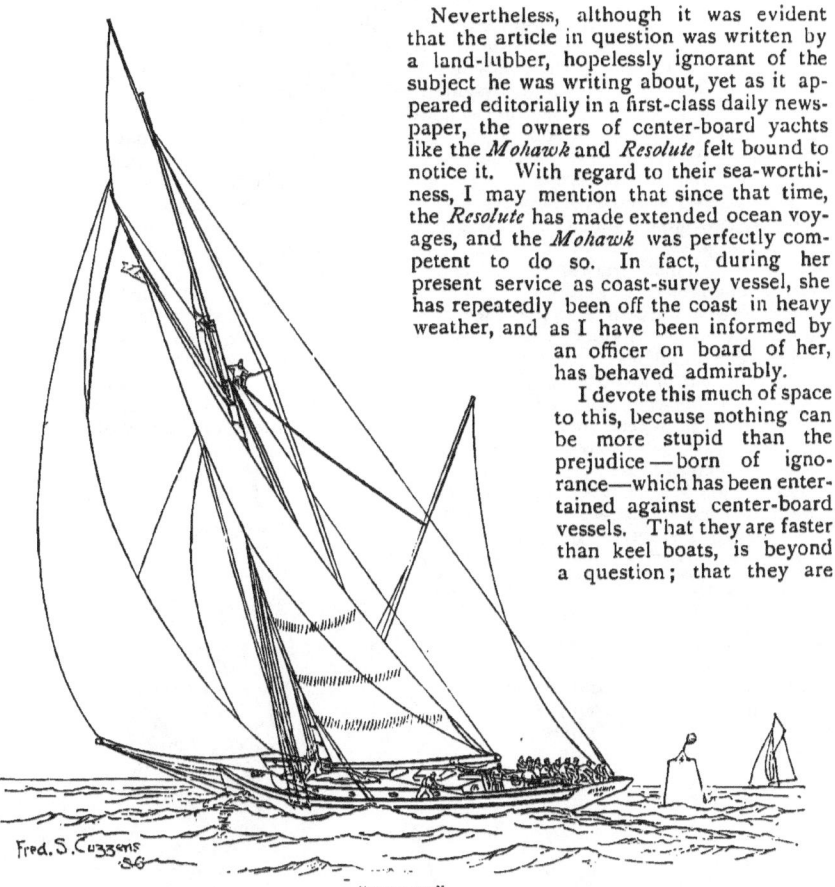

"MISCHIEF."

Nevertheless, although it was evident that the article in question was written by a land-lubber, hopelessly ignorant of the subject he was writing about, yet as it appeared editorially in a first-class daily newspaper, the owners of center-board yachts like the *Mohawk* and *Resolute* felt bound to notice it. With regard to their sea-worthiness, I may mention that since that time, the *Resolute* has made extended ocean voyages, and the *Mohawk* was perfectly competent to do so. In fact, during her present service as coast-survey vessel, she has repeatedly been off the coast in heavy weather, and as I have been informed by an officer on board of her, has behaved admirably.

I devote this much of space to this, because nothing can be more stupid than the prejudice—born of ignorance—which has been entertained against center-board vessels. That they are faster than keel boats, is beyond a question; that they are occurred to the competitors of the schooner *Livonia*, was due to the slight manner in which they were rigged, strength being sacrificed to neatness; and that the *Livonia's* exemption from damage was due to her strong and clumsy rig, and that the question of keel and center-board had nothing on earth to do with the matter. Moreover, as this gentleman should have known, two of the *Livonia's* opponents, *Sappho* and the *Dauntless*, were keel yachts. The idea of the *Mohawk* or any other center-board, "making leeway" is too ridiculous for mention, and the writer, probably, was not aware that the *Madeleine* with which he compares the *Mohawk* to the disadvantage of the latter, was also a center-board and quite as much of a skimming-dish model as the *Mohawk* was.

handier under canvas and better suited to our shallow harbors, cannot be doubted; and as to the question of safety, the percentage of accident in center-board craft is so small, that it need not be taken into account at all.

Before any of these challenge races were sailed, the New York Yacht Club had a fall regatta for cups offered by the then Commodore J. Nicholson Kane, and in addition, the sloop *Madcap*, a second-class yacht, challenged the *Vision* for the Bennett Cup. The difference in the sizes of the yachts was very marked, the *Vision* measuring 2,545 cubic feet, and the *Madcap*, 1,491 cubic feet, and receiving an allowance from the *Vision* of 12m. 44s.

Of the race I may say it was one of the most remarkable in the history of the club.

Of the sixteen yachts which started, only nine were able to get outside of the Hook. The others were caught by the young flood, and having no wind, could not stem it. Outside there was a cracking breeze and quite a heavy sea. The schooner *Peerless* was totally dismasted, and lost bowsprit, and the only ones which were able to get out to the light-ship were the schooners *Comet, Estelle*, and *Atalanta*, and the sloop *Sadie*.

The *Atalanta* was the only first-class schooner which made the course; all the rest were detained inside the Hook. Not a first-class sloop went the course, so the other prize winners were the schooner *Comet* and sloop *Sadie*. The latter was a deep Herreshoff production, and is now the schooner *Lotus*.

The first of what may be called the Hatch series of races came off October 6, 1875, and was a match over the New York Yacht Club course, between the schooners *Resolute* and *Estelle*, both center-boards. The difference in size between them was very marked, the entries being as follows:

NAME.	OWNER.	CUBIC FEET.	ALLOWANCE.
Resolute	Rufus Hatch	10,860	m. s.
Estelle	James D. Smith.	5,736	12 10

The race was a memorable one, the wind being fresh from east-south-east. The *Resolute* sailed in cruising trim, with boats at davits, anchors on bow, etc. The *Estelle* was in full racing trim, and was sailed by Capt. "Joe" Ellsworth; the *Resolute* being handled by her regular captain. To buoy No. 10 both yachts carried working top-sails, the wind in the bay being well to the eastward; and to this mark the *Resolute* had the best of the match, passing the buoy nearly four minutes ahead, the start being nearly an even one. Top-sails and flying-jibs had to come in, off the point of the Hook, and they began the beat to the light-ship under whole lower sails, the wind very strong from east-south-east, and the sea heavy. After getting outside, both went off for a long board on the port tack, the *Resolute* increasing her lead very materially. On the starboard reaches, the *Resolute* tacked for the light-ship nearly twenty-three minutes before the *Estelle* did, being

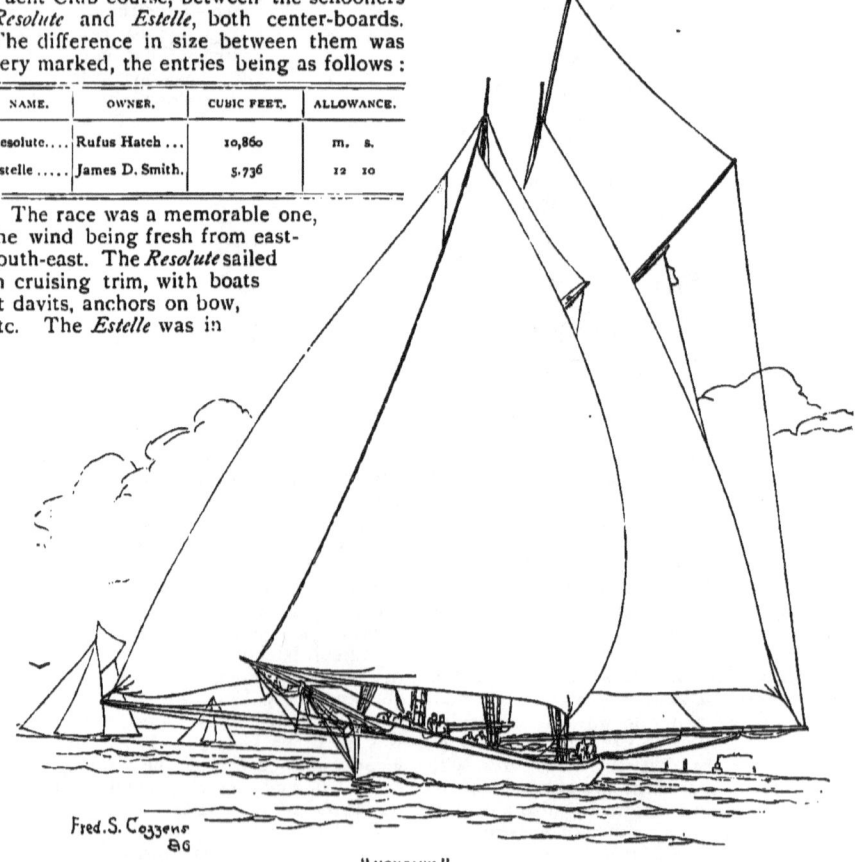

"MONTAUK."

at this point much more than her time ahead of the smaller yacht; and could she have weathered the mark on the port tack, the race had been hers beyond a peradventure; but the Captain deemed the risk too great, and had to make a short board on the starboard tack again. This necessitated going about twice, and this lost her the race, for, in the heavy sea, she was sluggish in stays, and lost much time; each time stopping when head to the wind, and making a stern-board. The *Estelle*, having the leading boat for a guide, stood far enough on her first reach on the starboard tack to weather the mark on the opposite one. The times of turning the light-ship were: *Resolute*, 12h. 34m. 30s.; *Estelle*, 12h. 39m.

Both yachts gybed around, with peaks dropped, and setting working top-sails, began the run in, wing and wing, with foreboom on the starboard side, and the *Resolute* also set her jib top-sail. As she ran in, a sea caught her under the counter, and she swung to, the foresail catching aback, and parting the guy; and as it went over it took the fore topmast out of her. She was, however, 6m. ahead at buoy No. 10, and finished 10m. 25s. ahead, not enough, however, to save her time, *Estelle* winning by 2m. 8s. Had the *Resolute* stood three minutes longer on her first starboard reach, before tacking for the light-ship, or if her captain had had a trifle more nerve, and squeezed her by that mark without tacking, she would have won the match with all ease.

The second of the Hatch series was sailed with the schooner *Comet*, October 8, and the disproportion of size between these yachts was greater than in the previous race, the entry standing:

NAME.	OWNER.	CUBIC FEET.	ALLOWANCES.
Resolute	Rufus Hatch	10,860	
Comet	W. H. Langley	4,662	17 38

The *Comet* was sailed by Capt. "Joe" Ellsworth, and the *Resolute* by her own captain. The wind was extremely light at the start, and Mr. Hatch desired a postponement; but Mr. Langley, knowing that the chances of his boat were better in light winds, refused to put off the start until another day; and not only that, but secured from Mr. Hatch a waiver of the eight hours' time limit, thus getting decidedly the best of "Uncle Rufus," as Mr. Hatch was popularly called. The race was a mere drift from start to finish. The *Comet* got out to the light-ship nearly three minutes ahead, and passed No. 10 buoy, on the return, over eleven minutes ahead. The *Resolute* overhauled and passed her before the finish line was reached, and went in ahead by thirty seconds; but of course the *Comet* won on allowance. Both yachts, however, were over the eight hours, so that if Mr. Hatch had not waived this condition, the race must have been resailed.

The third of the Hatch series of races was sailed October 12, and was a match of the *Resolute* against the *Dreadnought* and *Vesta* from the Sandy Hook Light-ship to Cape May Light-ship and return. There was no great difference in the size of the yachts, the *Vesta* and *Resolute* being centerboards, and the *Dreadnought* a keel boat. They had a splendid run to the Cape May mark, the *Resolute* turning over ten minutes ahead of the *Dreadnought*, and over twelve minutes ahead of the *Vesta*. On the return, the wind was unsteady and light, but the *Resolute* preserved her lead clear up to Barnegat, being then full eight miles ahead of the *Dreadnought*, the *Vesta* out of sight astern. From here to the finish, however, the *Resolute* had scarce any wind, and the *Dreadnought*, being farther off shore, had a trifle of air, and got to the line twenty-three seconds ahead, a winner, on actual time, of eight seconds.

The match was to be sailed according to club rules, and the allowance of the *Resolute* for 212 miles would have been 8m. 45s., which of course would have made her a winner; but Mr. C. J. Osborn, who owned the *Dreadnought*, insisted that the rule of the club as to allowance applied only to the New York club course, and that for the Bennett Challenge Cup to Cape May there was no allowance. Mr. J. D. Smith, the referee, so decided, and "Uncle Rufus" was once more deprived of a prize that, but for extreme bad luck, his yacht would have won.

The concluding race of the Hatch series, and the final contest for the year, was sailed October 28, and it was quite time to end the season, the days having grown very short, and the weather cold. The match was between the schooners *Resolute* and *Dauntless*, from off the club-house, at Stapleton, Staten Island, to and around the Cape May Light-ship, returning and finishing at the Sandy Hook light-ship. There is little to tell of the race. The *Dauntless* took the lead at the start, and increased it constantly to the finish, beating her opponent 1h. 52m. 18s., and making the race in 18h. 28m. 3s., beating the record; the best time before that having been 25h. 6m., and that with start and finish at the light-ship. The *Dauntless'* time has not since been beaten.

There had been an ocean race previous to this, viz., on October 26, between the schooners *Mohawk* and *Dauntless*, which grew out of the challenge of Mr. Garner, the owner of the *Mohawk*, he offering to sail any yacht, keel or center-board, twenty miles to windward from the Sandy Hook light-ship. As has been stated, the challenge was at once accepted by Mr. Bennett, whose yacht, the *Dauntless*, had, for a couple of years, been laid up at South Brooklyn. She was hastily prepared, and put in commission for this contest, and appeared at the starting line in very bad form. Her top sides had opened during the time she had been exposed to the sun while lying at the dock, and during most of the weather work to the outer mark she had considerable water in her lee bilge.

The *Mohawk*, under these circumstances, beat her easily, to windward; but, owing to some errors in sailing the *Mohawk*, she turned the outer mark only three minutes and forty seconds ahead. The error was in making three or four short tacks, permitting the *Dauntless* to go off by herself on a long board to the southward. Every time that the *Mohawk* tacked in a sea which was quite heavy, she of course lost something, and she took the risk of a shift of wind to the southward, which, if it had occurred, would have enabled the *Dauntless* to round the mark ahead.

After passing the mark, and leaving it on the starboard hand, instead of keeping off at once for the finish line, the *Mohawk* held her luff until she had passed to windward of the *Dauntless*, which was approaching on the starboard tack, and then kept off across her stern. This was for a bit of bravado, the captain of the *Mohawk* not doubting for an instant that, at running, his boat would be by all odds the best. In this he was mistaken, the broad, flat-bottomed *Mohawk* offering more of what is called " skin resistance " than did her narrow and deep opponent with her smooth copper bottom.

Both yachts, however, went in at a tremendous pace, the steam tug *Cyclops*, at that time the fastest in the harbor, being unable to keep up with them. She followed after the *Mohawk*, which, for some unaccountable reason, steered in north-west-by-west-half-west, although the course given for the outer mark had been east-south-east. The *Dauntless*, deceived somewhat by the courses steered by the *Mohawk* and the tug, went in north-west-by-west, three-quarters-west; but she too fell in far to the northward of the light-ship.

The *Mohawk* went right by the ship, without seeing her lights, and they were first seen by those on the tug, which at once hauled up for them, indicating to the *Dauntless*, by her whistle, the error of her course. She had been running wing-and-wing, with fore-boom to port, and had got well by the mark. Hauling to suddenly, her foresail gybed over, and the gaff, striking the triantic stay, was broken. All flying-kites were let go by the run, and the yacht was hauled sharp by the wind, and fetched in just to leeward of the mark, having to make a short board on the starboard tack, to weather it.

Of course the prize, a $1,000 cup, was hers; but had the *Mohawk* been sailed with better judgment in this race, she would have beaten the *Dauntless* by at least ten minutes. The next was the Centennial year, an important year for yachting, as for all other sports, and I may well reserve its events for the next article. As showing the wonderful development of the sport of yachting, I may say, in closing this chapter, that in the New York, the Brooklyn, the Atlantic, and Seawanhaka Yacht Clubs alone, there were sailed, during this year 1875, twenty-two races.

THE HISTORY OF AMERICAN YACHTING.

BY CAPTAIN R. F. COFFIN,

Author of "OLD SAILOR YARNS," "THE AMERICA'S CUP," etc., etc.

V.

FROM 1876 TO 1878.

1876, the Centennial of the United States, was a year of jubilee, and all out-of-door sports were immensely stimulated. Of course, yachting shared in the general prosperity, but in addition to the natural stimulus of the time of holiday, yachting had the extra excitement of an international contest, which, as we have recently, in 1885 and 1886, been made aware, is sufficient of itself to cause quite a yachting furore.

At the very beginning of the year — in January — and before anything had been done to release the pleasure fleet from its winter's seclusion, rumors began to be current that during this year there would come another challenger for the *America's* Cup. There did not seem to be any definite basis for the rumors, but they were floating around.

At a meeting of the New York Yacht Club, February 3, in this year, among the new members elected were Count Edward Batthyany, rear commodore of the Royal Albert Club, and Prince Maffeo Sciarra, of the Royal Italian Yacht Club, the owner of the schooner *Sappho;* and this gave rise to a rumor that the prince was intending to challenge with that vessel, and as was well known at that time, there was nothing in America that could beat her.

Rumors came also of a schooner building at Coburg, Ont., by a Captain Cuthbert, whose reputation as a builder of fast yachts was in Canada very prominent. He had, so it was reported, built a vessel named *Annie Cuthbert*, which had vanquished the *Cora*, one of Mr. McGichan's productions, and it was believed in Canada that he had only to build his schooner, send his challenge, and come here and take the cup.

It was also current gossip that a British gentleman, undeterred by the experience of Mr. Ashbury, would bring a schooner here from England for the cup. As to a cutter coming for it, such a thing was not thought of in those days, although the cutter was then, as now, the representative British yacht. All these rumors, however, had but slight basis. Meantime, the Centennial Commissioners at Philadelphia, desiring to have yacht racing on their programme, and having no course fit for it very near to the Quaker City, decided to have the races here, and they placed the matter in the hands of the following gentlemen: George S. Kingsland, commodore of the New York Yacht Club, chairman; John S. Dickerson, commodore of the Brooklyn Yacht Club, secretary; John M. Forbes, commodore of the Eastern Yacht Club; W. L. Swan, commodore of the Seawanhaka Yacht Club; W. T. Garner, vice-commodore of the New York Yacht Club, and S. Nicholson Kane, rear-commodore of the New York Yacht Club. It will be noted that all of these gentlemen, except Messrs. Forbes and Swan, were prominent members of the New York Yacht Club. Mr. Dickerson, although commodore of the Brooklyn, owed allegiance principally to the New York club, and if I remember rightly, was a member of its House Committee. Things in the yachting world have changed since then, and ten years after that no national committee would be considered truly representative that did not include a member from the Atlantic and American clubs.

The New York, however, was still, as in 1876, the club of this country. A writer in one of the dailies, of March 31, of this year, alluding to the Eastern yachts, says: "There are clubs in and around Boston, with numerous yachts; but their fastest ones, with the exception of the *America*, are second-class vessels which have been purchased from the New York clubs."

This was correct in 1876, but in 1886 the Eastern Club of Boston has the *Fortuna*, probably the fastest keel schooner in the world; the *Mayflower* and *Puritan*, center-board sloops, that are superior to anything in the New York Yacht Club, and a fleet of smaller yachts that are unrivaled in their respective classes. While the conservative and eminently respectable New

York Yacht Club has been standing still, and living on its past, its sister organizations, the Atlantic, the Eastern, the American, the Seawanhaka Corinthian and the Larchmont, have been going ahead with spinnakers pulling. Some of them are fully abreast of it now in the number and character of their vessels, and unless the old club obtains some new blood from somewhere, these other clubs will outrank it in popular estimation. The Eastern club, in 1885, and the Atlantic and Seawanhaka clubs, in 1886, by their spirited action in defense of the *America's* Cup, gained immensely in popular favor, and notified the New York Yacht Club that it was no longer considered able to defend this trophy, the emblem of the yachting supremacy of the world.

The building of open yachts, sloop and cat-rigged, was immensely stimulated by the action of the Centennial Commissioners, it being understood that one of the races would be for this class of yachts in New York Bay. In due time, the challenge of Major Charles Gifford, owner of the Canadian schooner-yacht *Countess of Dufferin*, for the *America's* Cup, was received and considered at a meeting of the club, held April 20, and as has always been the practice when this cup has been challenged, the club unanimously decided to waive the six months' notice, and to sail on any day most convenient to the challenger. Also, if he desired to sail in July, it was decided to give him two races — one over the New York club course, and one outside, and in case a third was necessary, the course to be determined by lot. If, however, Major Gifford preferred sailing in August, he was invited to join in the club's annual cruise, and to sail one race over the Block Island course, one race twenty miles to windward, the third, if a third race was necessary, to be determined by lot.

Meantime, the Centennial Yachting Committee decided to have three regattas, on June 22, 23 and 26, the first day over the New York club course, for yachts of fifteen tons and over; the second, in New York Bay, for all yachts under fifteen tons, and

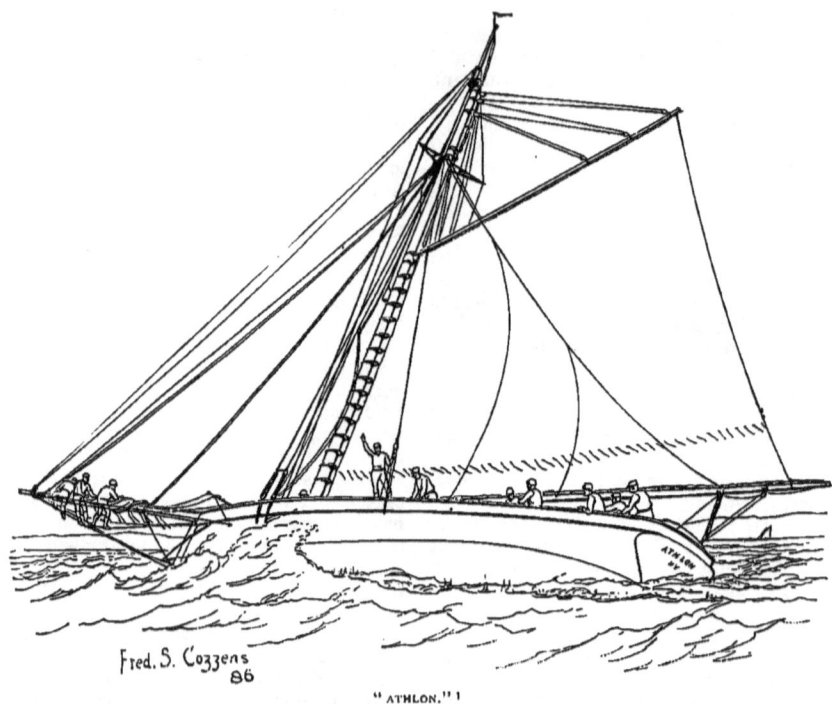

"ATHLON."[1]

[1] Sloop *Athlon*, owned by J. C. Barron, M.D., New York.

the third, a race from this port to and around the Cape May Light-ship and return. The prizes offered were the medal and diploma of the International Exhibition of 1876.

Beside these events thus early in the season provided for, there were, of course, the annual regattas of the clubs, the Brenton's Reef Challenge Cup race, fixed for July 22, and the Cape May Challenge Cup race, fixed for October. In preparation for these events, the schooners *Rambler*, *Dreadnought* and *Idler* were all lengthened this year, and many minor changes made in other yachts, the sail-makers being kept at work day and night.

I may mention just here that it was in this year that the first yacht at all approaching in model to what has, by common consent, come to be known as the cutter, was built, and that the designer of her, Mr. John Hyslop, who has contributed some interesting articles on yachting to OUTING, was considerably ridiculed, and was by some considered a trifle insane upon this subject of yacht designing. The yacht was called the *Petrel*, and she was 32 feet over all, 8 feet beam, 6 feet deep and 4 feet 6 inches draught. She was to have four tons of ballast, all of iron, inside.

It was in this year 1876, that the Seawanhaka club first came to New York from Oyster Bay, where it had been first organized,

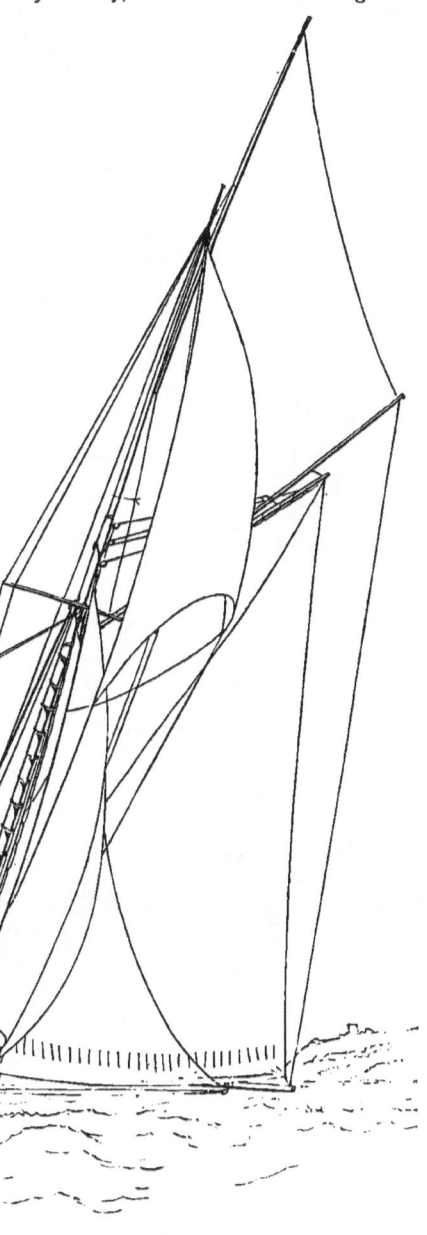

"MADGE."
Cutter *Madge*, owned by E. W. Sheldon, New York.

and it fixed on June 10 for a strictly Corinthian race, the course being the same as its present one, starting from off Fort Wadsworth and going around the Spit buoys to the light-ship. Always progressive, this club also arranged for a schooner race outside of the Hook, inviting entries from the New York, the Eastern, Brooklyn, Boston and Atlantic yacht clubs; yachts to be steered by owner or member of the club to which she belonged, but she could be manned by her regular crew. Meanwhile, a correspondence had been going on between Major Gifford and the Committee of the New York Yacht Club in relation to the proposed race for the cup, and finally, at a meeting held May 25, all of the and its largest sloop, the *Undine*, 52 feet, 9 inches mean length. It had two classes of sloops, four in each class.

We were hearing about this time much of the new Canadian schooner. Her trial trip had been a glorious success, etc. I can remember no trial trip of a yacht which has not been gloriously successful. They all sail well alone, and are tremendously fast with champagne accompaniment.

The New York Yacht Club, at its annual regatta, June 8, started a fine lot: three classes of schooners with four in the first, two and five in the third; and two classes of sloops, among them the *Arrow*, afterwards so celebrated. This being her first appearance in a regular regatta, although she had sailed with the fleet in the cruise of the club the previous year. Those who insist that we have made no advance in yacht designing, may be undeceived by the fact that the *Arrow*, at this regatta considered a marvel, would probably be beaten to-day by any modern sloop of her length. There was a strong breeze at this regatta, the *Gracie* and *Addie Voorhis* were obliged to withdraw, the *Arrow* and *Vindex* alone finishing in the first class of sloops, the *Arrow* beating the *Vindex* nearly 10m., winning class prize and also Bennett Challenge Cup prize. The schooner *Idler* made the best time on record over this course, viz., 3h. 54m. 48 1-2s. actual time, the start having been from off Stapleton, and the finish off buoy No. 15. The actual time of the schooner *Comet* in this race was 4h. 5m. 27 1-2s., but she took the Bennett Challenge Cup, her corrected time being 3h. 44m. 47 1-2s., showing that the cubical contents rule of the club for measurement for allowance of time was a perfectly fair one, enabling the smallest yacht in the schooner class to get in on even terms with one of the largest.

"SHADOW."[1]

Canadian gentleman's propositions were agreed to, and the races were fixed for the 10th, 12th and 14th of July, the club to name its yacht by July 1.

The Atlantic Yacht Club, now so strong and important, was in this year just beginning to come into notice. It started in its annual match this season four schooners, the largest, the *Ariel*, 72 feet mean length,

June 10, the Brooklyn Yacht Club and the Seawanhaka each had their regattas, the latter allowing entries only of sloops, and sailing with Corinthian crews. The Brooklyn event was notable for the finish having been in front of the new club-house in Gravesend Bay. I don't think it ever finished there after this year. It mustered however, three second class schooners —

[1] Sloop *Shadow*, owned by John Bryant, Boston.

the *Estelle*, *Comet* and *Gypsey*, four real good sloops in the first class, the *Gracie*, *Arrow*, *Undine* and *Kate* (now the *Whileaway*), with fair entries in the other two classes of sloops; in the lower class, the *Victoria*, *W. T. Lee* and *Susie S.*, open sloops of unrivaled speed. The *Arrow* beat the *Gracie* nearly 10m.

The entries for the first of the series of the Centennial races — that over the course of the New York Yacht Club, included eight schooners from New York, and one — the *Peerless* — from the Atlantic club. As Mr. J. R. Maxwell, owner of the *Peerless*, belonged also to the New York, it might be said that all of the schooners were from the old club. A similar race in 1886 would attract probably more schooners from either the Eastern or Atlantic clubs, than from the New York.

Of sloops, there were three from the New York, five from the Brooklyn, two from the Atlantic, and one — the *Schemer* — then owned by Mr. C. Smith Lee, from the Seawanhaka. The winners, I may mention, who captured the commissioners' medal and diploma, were the schooners *Dreadnought* and *Peerless*, and the sloops *Arrow* and *Orion*. They never got any other prize. There had been talk of valuable trophies in silver ware to be presented by the clubs, but so far as I remember, a "tarpaulin muster" scarce raised funds enough to pay the expenses of the committee.

There can be no doubt that the second day's racing was the event of the Centennial series; the owners of the open yachts were the only gentlemen that entered into the contest with the least enthusiasm. Owners of the large yachts had to be coaxed to start, but the men that had the small boats were eager for the fray, and cared for no prize other than the parchment of the commissioners. It will be a tolerable indication of the growing strength of the clubs, if I state the number of starters from each that came to the line, June 23. There were two from the Long Island club, five from the Williamsburgh, three from the Central Hudson, four from the Brooklyn, two from the Columbia, one from the Manhattan, four from the Pavonia, two from the Hudson River, one from the Seawanhaka, one from the Bayonne, one from the Mohican, two from the Jersey City, one from the Red Bank, one from the Perth Amboy, one from the Atlantic and one from the Providence Yacht Club; in all thirty-two yachts, many of them brand new.

The Providence entry was the famous catamaran *Amaryllis*, brought down to the city by the Herreshoffs. Some gentlemen who had seen this wonder sail, advised the owners of the second class boats to protest against her starting with them, but with calm confidence they replied, "Oh, let her come in, nothing can beat our sand-bag boats." So she started, and of course, beat the lot and could, I presume, had Mr. Herreshoff so minded, have gone twice over the course while the fastest of the sand-baggers made one circuit. After the race they protested, and curiously enough, the judges ruled her out. It made little difference to Mr. Herreshoff, however; he had introduced a new type of open yacht, and realized a favorite idea of yachtsmen for a half-century previous. It had always been a pet scheme with yachtsmen, that by a double hull, increased stability with a minimum of resistance could be secured; but it was not until Mr. Herreshoff applied the ball and socket joint, permitting each hull to accommodate itself to its own sea, that the speed was attained.

The *Amaryllis* has not had many successors, and this has seemed curious to me, for as an open yacht, the catamarans are superior to all others in every way. They are faster, safer, handier. They will not only sail fast; but they will lie still. There is one gentleman who has owned more of these craft than any one else, who is so expert in handling them, that he can do with them what cannot, without great risk, be done with any other description of open yachts; that is, weave in and out among the steamers and sailing craft of the most crowded part of the river front, and make a landing without the least damage. He has run side by side with the swiftest of the harbor steamers and beaten them, and has frequently gone the whole length of Long Island Sound with only a small boy as crew. Surely this cannot be done with any other description of open yacht. But ten years have passed since the *Amaryllis* came and conquered, and yet there are comparatively few catamarans, not above a score, I think, in the whole of the United States.

The third Centennial race came very near being an entire failure. Mr. Kingsland, owner of the schooner *Alarm*, being commodore of the New York club, and chairman of the committee, had to start her against the *America*, then recently purchased by General Butler, and the two sloops *Gracie* and *Arrow* were induced to start, and this was all. The *America* alone finished of the schooners, and the *Arrow* beat the *Gracie* 41m. 50s.; and so ended the series of Centennial regattas. Except the second, they were miserable failures.

Meantime, the cup committee of the club had named the schooner *Madeleine* as the defender of the cup against the challenging Canadian schooner, and there is no doubt that the choice was a wise one, for, on the whole, her record was the best. She had begun her career as a sloop, having been built by David Kirby at Rye,
She was at this time owned by Com. Jacob Voorhis, Jr., for whom she had originally been built, and who had expended much money on her in the effort to make her a success. He sold her, in 1875, to Com. John S. Dickerson, of the Brooklyn club, who owned her for many years. The only other yachts considered by the committee were the schooners *Palmer* and *Idler*, but there were no trial races, the qualities of each yacht being well known; and, as stated, the *Madeleine* was the final choice.

The arrival of the Canadian challenging schooner, *Countess of Dufferin*, was heralded by a great flourish of trumpets. The telegraph recorded her movements from Coburg to Quebec, and all the way down the St. Lawrence River, and at each reportable point passed by her until she arrived at this port. According to the highly seasoned reports, she was a flyer of most wonderful speed. "We raced two flying coasters, early this morning, for thirty miles, beating them hollow," wrote one correspondent, adding: "The sea-going qualities of our boat are fully established." "The yacht

"CROCODILE." [1]

Westchester County, in 1868, and had been successively altered, lengthened at each end and in the middle, and a second mast added, but never became at all famous for speed until 1873. This year, at Nyack, she was given a longer center-board, longer spars, and a new suit of canvas, and this season took position as queen of the fleet.
makes tremendous running," wrote another correspondent.

Meantime, in response to a request from Major Gifford, the cup races fixed for the 10th and 12th of July had been postponed to a later date.

I may mention here that on July 7, the old *America* had a narrow escape, having

[1] Sloop *Crocodile*, owned by J. G. Prague, New York.

struck on Brigantine Shoals, off the Jersey coast, and was taken off by the Coast Wrecking Company, in a leaky condition, requiring steam-pumps to keep her afloat. General Butler was on board of her. She was towed here and repaired.

The *Countess of Dufferin* arrived at New York July 17, and was found to have been a very poor copy of an American schooner yacht, and rough as a nutmeg-grater. The idea of putting a yacht like the *Madeleine* against her seemed absurd. There were scores of fishing schooners in this country more sightly than she, and doubtless more speedy. Her official certificate of measurement, from the Secretary of the Royal Canadian Yacht Club, stated her length at 91 feet 6 inches; beam, 23 feet 6 inches, and her tonnage at 200 tons.

The Brooklyn Yacht Club began its annual cruise this year from Glen Cove Harbor, July 20, and that it had begun to decline in importance was evident by the small muster of yachts for its annual cruise.

Notwithstanding the fact that its flagship had been chosen as a defender of the *America's* Cup, there were only present at this annual gathering the schooners *Madeleine*, *Clio* (at that time owned by the vice-commodore, J. R. Platt), *Tidal Wave*, *Sea Witch*, and *Mystic;* and of sloops, the *Niantic* (afterwards the *Hildegard*), *America* (afterwards the *Kelpie*), and *Favorite*.

It was while assembling in Glen Cove Harbor, on this occasion, that the news of the capsizing of the schooner yacht *Mohawk* was telegraphed to the club, and was at first discredited by all who were acquainted with the yacht. It was the almost universal opinion that the masts would have gone out of the yacht before she could have upset, but later intelligence confirmed the first announcement, and a gloom was thrown over the cruise at its very beginning. The particulars of this sad accident were that the *Mohawk* was lying off Stapleton, Staten Island, with all after canvas set, even to her enormous club top-sail. The owner, Mr. William T. Garner, was on board, with his wife and a few friends. The yacht was just getting under way, her chain had been hove short, and her jibs had been run up, in order that, as she gathered way, she might break out the anchor from its hold on the bottom, the capstan being of insufficient power. The helm was a-weather, when a hard squall from the north-west struck the yacht, as she lay without way, and without the possibility of gathering way, and she went down until she filled and sank. Mr. Garner was drowned while trying to rescue his wife from the cabin. Some ballast had shifted and pinned her fast, so that the effort was unsuccessful, and she also lost her life.

Much unmerited criticism was made upon the *Mohawk's* model, and upon center-board yachts generally, and a check was given to the sport from which it did not recover for years. In point of fact, the *Mohawk* was as safe a vessel as ever floated. She was lost through the grossest carelessness, and in consequence of the over-confidence felt in her stability. There has been no vessel yet built in this world that cannot be wrecked by careless handling, and that the *Mohawk* upset was in no wise due to any defect of model. Properly handled, she was more than ordinarily safe.

The third race for the Brenton's Reef Challenge Cup, afterwards happily carried away to Europe by the cutter *Genesta*, was sailed July 27 to 29, 1876. It was one of the four offered by Mr. Bennett when vice-commodore of the N.Y.Y.C., in 1872, the other three being the challenge cups for schooners and sloops over the regular course of the club, and the Cape May Challenge Cup, also captured later on by the British cutter *Genesta*.

This cup race had never been a popular one. It had been offered time and again, and no entries for the race had been received, and it had been sailed for but twice, each time by the schooners *Rambler* and *Madeleine*, one of the races being a return match growing out of the first. It would not have attracted any entries this year, had it not been that the owners of the *Wanderer* and *Idler* desired to give the Canadian visitor a chance. General Butler desired to exhibit the *America* as a winner, and the owner of the *Tidal Wave*, desiring to sell her, thought it would add to her value if she could win this race. Right here I may say that, if he had expended a few hundred dollars for new ropes and sails, it is very probable that she would have been the victor; certainly she would have had a long lead at the Brenton's Reef Light-ship, as she showed a better turn of speed on the reach along the coast than either of the other yachts, and although continually breaking down, she was the first to turn the mark.

The *Countess of Dufferin* did not enter for this race, but she started with the yachts "just to see what she could do," her owner said. In his heart of hearts, he believed he was going to come home at the end of the race far in advance of any other yacht.

of the secretary's certificate, already given. By Mr. Smith's tape-line, she was 100.85 feet over all, 95.55 feet water-line, 23.55 feet beam, 7¼ feet deep.

The Atlantic Yacht Club this year started from Glen Cove on its annual cruise July 30, and it did not muster a very large fleet, but it was larger than on previous years. There were the schooners *Triton, Peerless,* and *Agnes,* and the sloops *Undine, Orion, Nimbus, Myra, Genia, Lotus,* and *Hope.* There was a great gathering of open racing craft this year at Newburgh, the starters numbering twenty-five, among them being the *William R. Brown,* a new racer just built at Brooklyn by Mr. Harry Smedley, and out of this Newburgh race grew a series of two matches between this yacht and the *Susie S.,*

"ZOE."[1]

So the starters were the schooners *Idler, Wanderer, Tidal Wave, America,* and *Countess of Dufferin.* The *Idler* won, with the *Wanderer* second, the rest nowhere. When the *Idler* was sold, the cup reverted to the club, and was never again competed for until the schooner *Dauntless* and the cutter *Genesta* sailed for it, in 1885. In all probability it never would have been sailed for, for the owners did not like the course. They were, therefore, very willing that the British cutter should take it over, because it gives them an opportunity of a race in British waters, unhampered by British rules of measurement. For the races for this cup there is no allowance of time, and the beamy yacht can't be discriminated against. Just before the start on this race, the Canadian yacht was measured on the dock by Mr. A. Cary Smith, at that time measurer of the N.Y.Y.C., and he made her somewhat different from the measurement

for $500 a side in each race, one at Newburgh, and the other in New York Bay. I may add that only one was sailed — that at Newburgh, where the *Brown* belonged — and her owner paid half forfeit, and did not come down the river to be beaten.

August 3, Major Gifford, owner of the *Countess of Dufferin,* asked for a further delay of the races for the cup until August 14, but finally agreed to be ready August 11, and it was arranged that this should be the date of the first race, the other to be sailed August 13, and the third, if necessary, August 14.

The first race for the cup was sailed on the day appointed, the entry being:

NAME.	WATER-LINE LENGTH.	CUBIC CONTENTS, FEET.	ALLOWANCES.
Countess of Dufferin..	95.53	9,028.04	M. S. allows.
Madeleine	95.02	8,499.17	1.1

[1] Sloop *Zoe,* owned by H. A. Sanderson, New York. (One of the Larchmont cracks.)

The Canadian yacht showed much better in this first race than had been expected. But she had been much improved since her first arrival. Captain Cuthbert, her builder and sailing master, was a friend of the Elsworths, of Bayonne, N.J., expert yacht racers from their boyhood, and under their advice, the *Countess* had been placed upon the dock at Port Richmond, and scraped and sand-papered, and made as smooth as was possible, and she was then given a coat of pot-lead and tallow. All her sails, also, with a few exceptions, had been made in New York, and so, as a daily paper remarked, whichever way the contest terminated, it would be a victory for the American model.

The race, however, attracted much interest, although not a tithe of that evinced when the *Cambria* and *Livonia* came, or when the *Genesta* appeared as a challenger. Still there were a dozen excursion steamers and a couple of dozen of yachts present at the start ready to go over the course with the racers. The start was a pretty one, and that the reader may judge of the quality of the two yachts, the following table of times is given, the wind being south, a moderate breeze, and the tide last quarter flood:

NAME.	START. h. m. s.	BUOY 10. h. m. s.	LIGHT-SHIP. h. m. s.	BUOY 10. h. m. s.	FINISH. h. m. s.
Madeleine	11.16.31	1.19.19	2.51.52	3.57.28	4.41.26
Countess of Dufferin	11.17.06	1.26.32	2.56.33	4.06.48	4.51.59

The *Madeleine*, therefore, won by 9m. 58s. actual time, and by 10m. 59s. corrected time.

The second and concluding race was sailed the next day, and the course was twenty miles to windward from Sandy Hook and return, the wind light throughout from south-south-east, and the water smooth. The old-yacht *America*, the original winner of the cup, stripped for a contest, sending all weight ashore that could be spared, and in racing fettle went over the course with the other two yachts, beating the Canadian yacht, but being in turn beaten by the *Madeleine*; and as a matter of comparison, I will give the times of all three schooners over the course:

NAME.	START. h. m. s.	OUTER MARK. h. m. s.	FINISH. h. m. s.
Madeleine	12.17.24	5.01.52	7.37.11
Countess of Dufferin	12.17.58	5.13.41	8.03.58
America	12.22.(?)	5.04.53	7.49.00

It will be seen that to the outer mark, the *America* beat both the other schooners. The difference in time between the *Madeleine* and *Countess of Dufferin* at the finish does not accurately represent the distance between them, as after the *Madeleine* finished, the wind failed, and the tide being ahead, the *Countess of Dufferin* was a long time doing a short distance. She was, however, decidedly beaten.

August 14, the fleet of the New York club assembled in Glen Cove Harbor to begin the annual cruise. Of schooners, there were the *Alarm* (the flagship), *Restless*, *Wanderer*, *Dreadnought*, *Estelle*, *Rambler*, *Palmer*, *Idler*, *Foam*, *Vesta*, and *Meta*. Sloops *Arrow*, *Vindex*, *Vision*, *Windward*, and *Wayward*.

The programme agreed upon was an extended one, and included a visit to Greenport, Vineyard Haven, Marblehead, the Isle of Shoals, Portland, Provincetown, Vineyard Haven, and Newport, sailing races there August 28 and 29, and disbanding the 30th. All this was changed later on, and the new schedule was from Shelter Island to New London, Newport, and Vineyard Haven, and the participation in the regatta of the Eastern Yacht Club at Swampscott was abandoned. The cruise was a very tame one, and the fleet broke up at Edgartown August 21.

Probably if the Canadian schooner had accompanied the fleet, it would have taken her around Cape Cod, and the original programme would have been adhered to; but Major Gifford had had enough of it. The funds of the syndicate of club members which had built and sent her to the contest were running low, some new sails purchased in New York were yet to be paid for, and things were in no condition for a junketing excursion.

As it was, although the cruise was practically at an end at Vineyard Haven, some of the yachts went on to the east as far as Provincetown, returning to Newport August 27, where they finally separated.

It was about the last of August that Mr. J. E. Loubat, owner of the schooner-yacht *Enchantress*, presented to the New York Yacht Club a $1,000 cup to be sailed for October 12, open to all schooner-yachts of 100 tons and upward, belonging to any organized club in the world, on an allowance of twelve seconds to the ton; New York Yacht Club rules to govern in all other respects. The course to be from off Owl's Head, to and around Sandy Hook Light-ship, thence to and around the Cape

May Light-ship and return. I may mention that the only entries for this prize were the schooners *Idler*, owned by Mr. Samuel J. Colgate, and the *Atalanta*, owned by Mr. William Astor. The start was made October 12, and the *Atalanta* won, beating the *Idler* 3h. 10m. 3s. It was a fluky race, and the result did not correctly show the relative merits of the two yachts. It is because these long races have always been determined mainly by chance, that they have proven so distasteful to yacht owners.

This race concluded the racing of this remarkably active yachting year, but previous to this, on September 16, the Seawanhaka club had a fine fall regatta at Oyster Bay, and also a fall event over the regular club course on September 19. The entries to the latter event, however, were few, only two schooners in each of the two schooner classes, and a single sloop in each of the two classes of that rig, showing that owners had become tired of racing. The Seawanhaka club finally wound up its season by a Corinthian race for all second-class schooners over its New York course, and the Brooklyn club had also a concluding race over its regular course. The Atlantic club also had a pennant regatta on September 23. Beside these, there were fall events in all the minor clubs in this neighborhood, showing that the impetus given to the sport by the challenge for the *America's* Cup was felt to the end of the season.

Early in the year 1877, what may be called the second "cutter" ever built in this country, was begun by Mr. John Mumm, at the foot of Court street, Brooklyn, from a design by Mr. Robert Center.

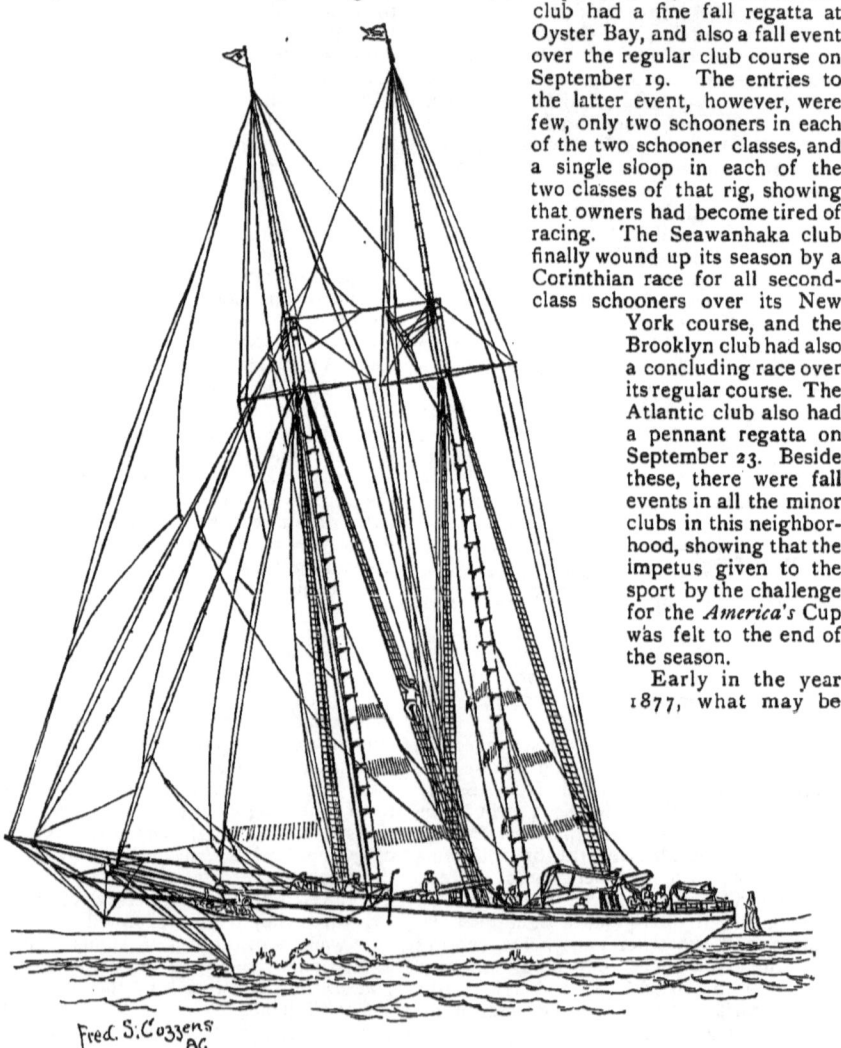

"GITANA."[1]

[1] Schooner *Gitana*, owned by Wm. F. Weld, Jr., Philadelphia.

She was 40 feet on the water-line; 12 feet beam, and 6 feet 10 inches deep, and she had a 1½ ton of lead outside. She was more like the British cutters in her model and rig than the *Petrel* had been, for her jib set flying, and the bowsprit was a sliding one. Later on, when what had been aptly called the "cutter craze" became virulent, there were writers who assumed all the credit of the introduction of yachts of this type into this country, but the fact is patent that Mr. Hyslop and Mr. Center were the first two gentlemen who brought practically to the notice of the American yachtsmen the British cutter, and claimed for it a superiority over the ordinary center-board sloop. The name of this keel-sloop or cutter was the *Volante*, and she proved a much greater success in the matter of speed than her designer ever believed possible, his only aim in her design being to produce something entirely safe for two young relatives to sail in.

I think it must be admitted that the Seawanhaka Yacht Club has done more to promote yacht racing than any other organization during the time of its existence. It has never aimed at being a social club, but always a racing one, and in March of this year it adopted its racing programme for the season, appropriating $900 for a Corinthian race June 16, for first and second class schooners open to all clubs. An Ocean race for first and second class schooners, June 23, owners to steer, open to all clubs, appropriating $585 for prizes; an annual regatta at Oyster Bay July 4, $425 for prizes; a race for open boats at Oyster Bay, July 28, $50 for each class; four races for open boats at Oyster Bay, the last four Saturdays in September, $50 for each class, and a Ladies' day in September, at an expense of $200. With the exception of the two last fixtures, all the races were sailed as arranged.

The Seawanhaka club also during this winter initiated a series of lectures on yacht designing etc., which have proved to be of immense benefit to its members, starting many of them on a quest for information in this direction, the result being apparent in a much better class of yachts in the succeeding decade.

The Brooklyn Yacht Club disposed of its house on Gravesend Bay this year, and took one more step backwards by not providing itself with another. I think it may safely be assumed that if a yacht club has no head-quarters and anchorage, it will drop astern of its sister organizations.

The New York club may perhaps be cited as an exception to this rule, but that club has not progressed as it should have done, since its house on Staten Island and its anchorage off Stapleton were given up, and, therefore, the rule holds good even in regard to that organization.

Meantime, yachting in the harbors of the New England States had been making great advances. The total number of American yacht clubs in 1877 was fifty-three, of which number twelve were in the New England States and mostly in Boston and its neighborhood. In and around New York City there were twenty-three. There were eight on the Lakes, and ten in Southern waters. In the aggregate membership were 772 owners.

Among the New England clubs, the Eastern was in 1877, as it is in 1886, the principal, and it then had twenty-nine schooners, twelve sloops and two steam yachts on its rolls, but it must be noted that of their schooners, all the large ones were New York rather than Eastern club yachts. For instance, there were among them the *Dauntless*, *Alarm*, *Columbia*, *Faustine*, *Enchantress* etc. It had an aggregate of 233 members, but as with the yachts, many of them owed their principal allegiance to the New York Yacht Club. It may be interesting to know that the commodore of the Eastern Yacht Club, in 1877, was Mr. J. Malcolm Forbes, who in 1886 owned the celebrated sloop yacht *Puritan*.

Next to the Eastern in importance, and its senior in age, was the Boston club with 258 members, who were all, or nearly all, Boston yacht club men. Its muster roll comprised seventy-eight yachts, of which fourteen were schooners, and sixty-one sloops, with three steamers.

The Dorchester club, like the Seawanhaka of New York, was from the first a racing organization. It had averaged six races each season since 1870, when it was first organized.

The South Boston club had about thirty yachts in 1877 and a membership of 150.

The Beverly club had ninety-six members and fifty-four yachts, mostly small, open cat-rigged affairs, handled almost invariably by their owners, and requiring more skill than any other class of yacht that can be named.

The East Boston club organized in 1874 had in 1877 twenty-five members.

The Portland club in 1877 had 140 members and twenty-five yachts.

The Bunker Hill club, organized in 1869, had in 1877 twenty-six yachts, nine of which were schooners, the largest of which, however, was but 39 feet 9 inches in length, and the smallest 18 feet.

The New Bedford club was organized in 1877, and as is well known to yachting men, has been one of the most prosperous in the country since that time.

The Lynn club, organized in 1870, with eleven yachts and sixteen members, had in 1877 thirty-seven yachts and 132 members.

The Haverhill club, organized in 1874, with about a dozen members and some half dozen yachts, had in three years grown to thirty members and thirteen yachts, two of which were steamers.

The Quincy club, also organized in 1874, had grown to be an active and flourishing organization in 1877, and has since that time gone ahead with a spinnaker breeze.

The above bare mention of the New England clubs, will show how the sport of yachting had broadened out. Each of these clubs had at least one, and some six and eight races during the season, and builders and sail-makers in Boston were kept busy for the whole year round.

1877, however, was a dull yachting year. It was the natural reaction from the animation and excitement of the centennial year which preceded it, and at its beginning, there were only three new yachts building in all the United States, if the open boats be excepted; and of these, there were much fewer than usual.

A feature of yachting in 1877 was the building of the double-hulled schooner yacht *Nereid*, for Mr. Anson Phelps Stokes, at Staten Island, by Mr. "Lew" Towne. She came afterwards to be known as *Stokes' Folly*, but when first built, she frightened the owners of second class schooners so much that a special meeting of the New York Yacht Club was held, in order to take measures to bar her out of the races, and the movement came very near succeeding. The hulls were three feet wide and placed ten feet apart, she was schooner rigged, with masts 43 feet and topmast 20 feet, boom 28 feet and gaff 14 feet, the hulls were 5 feet deep and each had a 5 foot centerboard. She was steered with one rudder hung between the two hulls; in one hull were accommodated the officers and crew, and in the other the owner and guests. Not having the Herreshoff ball and socket joints, and the connections between the hulls being rigid, she was an entire failure.

Mr. Charles A. Meigs, of Staten Island, was quite an enthusiast on the subject of double-hulled yachts, and he had one built this year at the foot of Court Street, Brooklyn. The hulls were 46 feet each, of 3 feet 6 inch beam. The connection between them was rigid, and of course the usual result, flat failure, followed. Any one looking at a Herreshoff catamaran as she bounds along, each hull having its own independent motion, will realize what the hulls of the boat rigidly connected desire to do and are unable; one of two things must surely happen, either there will be no speed or the connection will break, to permit the desired motion. In the case of Mr. Meigs' boat, she was built in the most flimsy manner, and went all to pieces.

I ought not to omit to mention, in a history of American yachting, that the sport of racing miniature yachts attained quite a prominence during the years 1876 and 1877, the principal head-quarters for this sport being the lake in Prospect Park. The lessons learned there have been apparent in many changes in model and rig adopted since. It would have been well if the sport had been encouraged, but after a year or two it fell into disuse.

The annual June regatta of the New York Yacht Club, this year, was memorable for the squall at the finish, which caught the schooners *Rambler* and *Wanderer*, with all sail set, and obliged them to let everything go by the run, the *Wanderer* forging over the line a winner, with her club topsail flying far out to leeward like an immense flag of triumph, and her balloon main topmast staysail dragging under the lee counter. The *Rambler*, in a hardly less disheveled condition, was but one minute behind her. The scene is well portrayed in a picture by A. Cary Smith, a copy of which is to be seen in all collections of yachting pictures.

All the clubs had their annual events as usual, but for the most part they were tamer affairs than usual, and there was nothing especial to note about them. Perhaps the most noteworthy was the Corinthian match of the Seawanhaka club, which mustered two second-class schooners, two first-class sloops, and nine second-class sloops. There was a fresh breeze, and during a portion of the race, a hard rain, the amateurs doing their work with all the efficiency of professional seamen.

June 22, a novel accident occurred during a race between the catamarans *Amaryllis* and *John Gilpin*, both Herreshoff boats, the former, the original one, introduced at the Centennial regatta the preceding year ; while going at a very rapid rate, the bows of the two hulls ran under, and her momentum was so great that she turned completely, end over end. Since that time, the hulls have been built with a rank sheer forward, in order to counteract this tendency to run under.

The cruise of the Brooklyn Yacht club this year was a miserable failure, only seven yachts putting in an appearance at the start, which was diminished to five at the close of the cruise, two schooners and three sloops.

The New York Yacht club, however, had a fine muster of yachts, and left Glen Cove, August 8, visiting New London, Greenport, Block Island, Vineyard Haven, New Bedford and Newport, disbanding there August 17.

A race for the Bennett Cape May Challenge Cup, and the last contest for this prize previous to its being captured by the British cutter *Genesta*, was sailed September 4 to 6 ; the starters having been the schooners *Idler*, *Rambler*, *Vesta* and *Dreadnought*. The *Idler* was the winner with the *Rambler* second. When the *Idler* was sold, the cup came back to the club, which held it until 1885, not having been able to obtain any entries for it, although it was frequently offered and days set for the race. As an interesting incident of this year's yachting, I may mention the launch of Mr. William Astor's schooner *Ambassadress*, the largest sailing yacht ever built in this country. She was built by Mr. David Carll, at City Island, and launched September 19, and is 148 feet long, 29 feet beam, 12 feet 3 inches deep, and 11 feet draught.

September 27, the Atlantic Yacht Club's fall regatta was sailed with three schooners and seven sloops as starters. Most of the time during the race the fog was extremely dense, and on the return off Sandy Hook, the committee's tug, *Cyclops*, ran into the Richmond steamer, *Isaac Bell*, damaging her seriously, so that she had to return to the city for repairs. I close the record of the year 1877, and this article, already too long, by recording that on November 9, the schooner yacht *Ariel*, a sister vessel to the *Clio*, started on a voyage to San Francisco, having been purchased by a gentleman there, and that she arrived there all right in due time, proving once more, if any proof was needed, that center-board yachts, even of the smallest size, can safely make an ocean voyage.

THE HISTORY OF AMERICAN YACHTING.

BY CAPTAIN R. F. COFFIN,

Author of "Old Sailor Yarns," "The America's Cup," etc., etc.

VI.

FROM 1878 TO 1885.

VERY early in the year 1878, Mr. Lester Wallack, the celebrated actor, at that time the owner of the famous schooner *Columbia*, was elected Commodore of the Brooklyn Yacht Club, and he was, I think, its last commodore for many years. In a little speech which he made on assuming his office, Mr. Wallack frankly confessed that he was no great sailor and no great yachtsman. He was, as all know, a very estimable gentleman, but about the most unsuitable person that the club could have selected, in view of its waning fortune, to take the executive charge.

I may mention, as something which has had a decided influence for good on American yachting, that during the winter of 1878, Mr. A. Cary Smith, by invitation of the Seawanhaka Yacht Club, delivered a series of lectures before its members at Delmonico's, on Naval Architecture. The information thus obtained has been supplemented by study in other quarters, and the result has been the introduction of a better class of yachts, and more perfectly fitted, than before. It was in the early part of 1878, that the keel schooner *Intrepid* was built at Brooklyn by the Poillons, from a design by Mr. A. Cary Smith. While upon the stocks she was very extensively criticised. It was asserted that she was too fine forward, her "dead flat" too far aft, that she would bury in driving hard, etc. She falsified the predictions of these wise people, by proving a success in every way, and was one of the finest yachts in the fleet. Her owner, Mr. Lloyd Phenix, being an expert navigator, has made several foreign cruises in her.

In May, 1878, the schooner yacht *Mohawk* was sold to the United States Coast Survey Service, and her name changed to the *Eagre*. It is notable that after a year of more than ordinary excitement, such as occurs always, when an international event is one of the season's incidents, the next year is marked by a general dullness and this was particularly the case during the season of 1878. The clubs all had their regattas as usual, but they were tame affairs, the entries few and the attendance small. The New York club tried to have a race June 13, but it failed from lack of wind, and was sailed June 14, in the presence of only the committee and a few reporters. There started two keel schooners, two first-class and three second-class centerboard schooners, only one first-class sloop, the *Vision*, and four second-class sloops.

A notable race of this season, was a contest of small open yachts in the bay. The affair was organized by a volunteer committee of gentlemen interested in yachting, the money for the expense being obtained by subscription and the entry made free. It drew together forty-three starters, divided into five classes, and was an extremely successful affair.

In July of this year, the cutter *Muriel* was built for Mr. James Stillman, by Mr. Henry Piepgras at Brooklyn, from a design by Mr. John Harvey of England, this being the first real *bona fide* British cutter ever built in this country. She was 45 feet over all; 9 feet beam; 6 feet, 3 inches deep; 7 feet, 9 inches draught, and carried 6½ tons of outside lead. What came to be called the "cutter controversy" was just then beginning to rage in this country, and the advocates of the British boat were claiming superior speed for their favorite model, which was as strenuously denied by the centerboard partisans, and Mr. James Stillman, a prominent member of the New York Yacht Club, then the owner of the schooner yacht *Wanderer*, determined to test the question practically by having a yacht built from the lines as near as might be of the fastest of her class in Great Britain. The *Muriel* was not a success in the matter of speed, nor have any of the successors of this type been, the centerboard boat, in good breezes, having always proven the most speedy. It has also been proven, that this style of yacht is less comfortable than the broad centerboard, and not suited for the shallow American harbors. They are, however, very handsome craft, and out of the controversy as to cutter and centerboard, has come a compromise between the two extremes of broad and shallow and deep and narrow, which is superior to either. The centerboard is retained, but with it is a keel, through which it plays. The yacht is made narrower and deeper than of old, the lack of stability due to narrowing the model, being made up by outside lead.

The *Muriel*, however, attracted much attention, and considerable ridicule when she first appeared. The Seawanhaka club was first to lead off this season with a cruise; the first Corinthian cruise ever attempted in this country; the yachts being all manned and sailed by amateurs. The fleet started from Oyster Bay, L. I., and it consisted of one schooner and six sloops. It went on to New London and thence to Newport.

The Atlantic Club was the next to start a fleet, and had six schooners and twelve sloops, and it signalized its cruise by giving a regatta at

always been a favorite stopping-place for this club, and at one time it contemplated making this port its headquarters. Fortunately, the project fell through.

The New York Yacht Club mustered ten schooners and four sloops for the annual cruise, and went direct from Glen Cove to Greenport, getting there while the fleet of the Atlantic Club was in the harbor. It

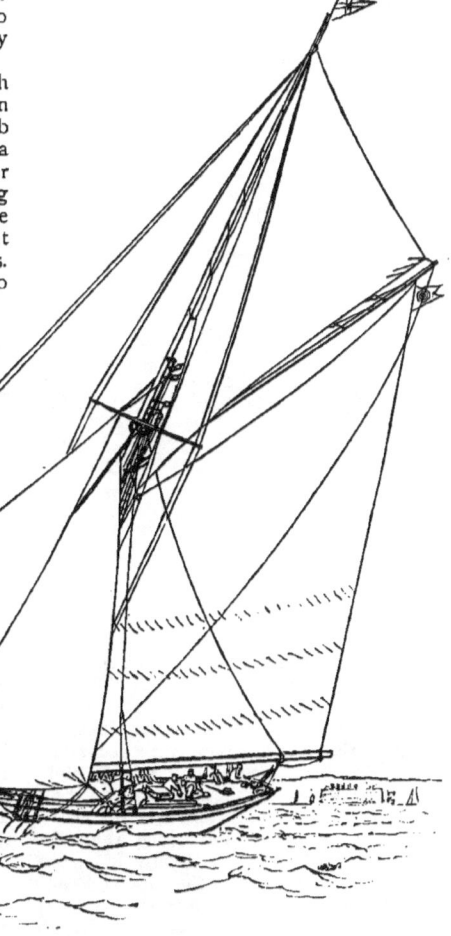

Cutter "Bedouin." Owned by Mr. Archibald Rogers, New York.

Greenport, L. I., starting twenty-six boats, twelve of which did not belong to the club. These regattas in Greenport were features of the Atlantic Club's annual cruises for several years. It has went from Greenport to New London, thence to Newport, and thence to New Bedford, where a race was arranged for the purpose of giving the Boston sloop *Thistle* an opportunity to test her speed with the

New York sloops. The race was sailed August 14, the *Thistle* sailing against the *Active*, *Vixen* and *Regina*. The Boston yacht started ahead and led all around the course, but was beaten by the *Vixen*, 1m. 14s. elapsed, and 2m. 57s. corrected time. She was miserably sailed, however, and it was my opinion at the time, I having been on board of her during the race, that had she been as well handled as the New York sloops, she would have beaten them. The *Active* beat the *Thistle* 27s. and the *Thistle* beat the *Regina* 2m. 44s. The course was twenty miles.

I may mention as an incident of this cruise, that in a run from Vineyard mark. They tried again October 22, and made the race, the weather having been moderate and sea smooth, and the *Gracie* won by 13m. 46s., thus ending the season of 1878.

The next season was a dull one and there was little of note in its events. The

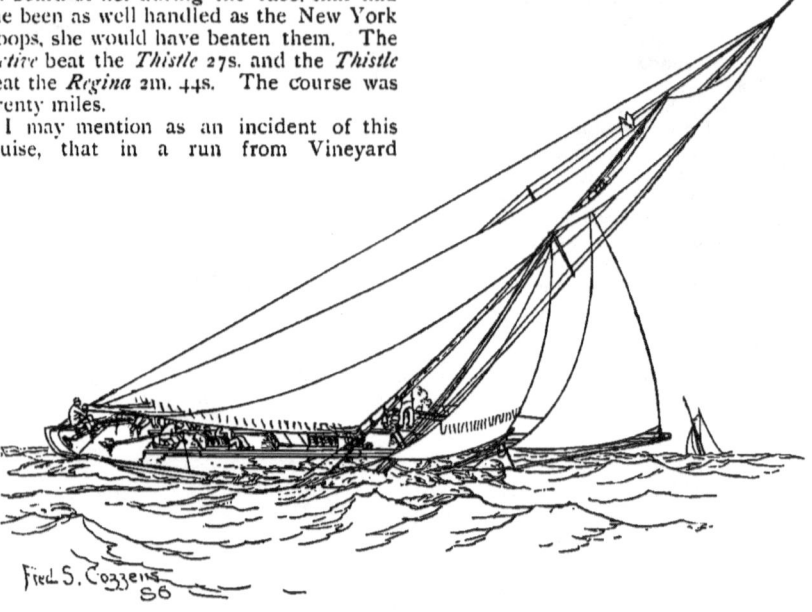

¹ "MAGGIE."

Haven to Newport, the double-hulled schooner *Neried* beat the fleet, gaining her only victory. She took a short cut through Woods Hole, gaining a fair tide thereby, and arrived at Newport twenty seconds ahead of the *Vixen*, which came second.

The Brooklyn club issued a most elaborate programme for a cruise, but no yachts appeared at the rendezvous and the cruise did not take place, and since then the Brooklyn has been a club only in name. October 15 of this year, the sloops *Gracie* and *Vision* attempted a race twenty miles to windward from the Sandy Hook Lightship. The *Vision* was of the most pronounced skimming-dish type, drawing but 4 feet, 10 inches of water on a water-line length of 60 feet, 2 inches. The *Gracie* drew 6 feet, 3 inches on a water line of 65 feet. Neither was fit for ocean racing, and both were disabled and failed to reach the outer

clubs, big and little, had their regattas, the entries few and the interest trifling, and confined altogether to the particular club whose yachts were racing.

It was in June of this year, that Mr. Piepgras built the cutter *Yolande*, the second real British cutter ever built in this country. She was built in the yard attached to Mr. Piepgras' dwelling, and then moved through the street to the water, several blocks distant.

I call her a cutter, because by common consent this name has been given to deep, narrow yachts, similar in model and rig to the one-masted vessels common in England, and to distinguish them from the broad and shallow centerboard sloops. Of course, properly speaking, the rig should govern the designation in this, as in all other craft, ship, bark, brig, schooner, etc.; but we needed some appellation which should

¹ Cutter "Maggie." Owned by Mr. L. Cass Ledyard. New York.

designate the shape of the hull, and this term "cutter" has been adopted. The *Yolande* was a cutter to all intents and purposes; cutter in model and cutter in rig. So anxious, however, have the advocates of English yachts been to prove that the cutter could beat all creation, that every sloop which has proven at all fast, has been dubbed a cutter, and the term has become rather confusing. I intend, when speaking of cutters, to designate such yachts as the *Muriel*, *Yolande*, *Bedouin*, *Wenonah*, *Stranger*, *Madge*, *Clara*, etc., and not such as the *Huron*, *Thetis*, *Puritan*, *Mayflower*, etc.

The *Yolande* was built for Mr. M. Roosevelt Schuyler, the most pronounced advocate of the cutter model that we have ever had in this country. Mr. Schuyler was an extremist; not only did he believe the cutter possessed of superior excellence, but he insisted that all other types were faulty in the extreme and could have no good quality. The *Yolande* was 32 feet over all, 25 feet water line, 7 feet, 6 inches beam, and 5 feet deep. She had a deep rocker keel composed entirely of lead which weighed 8,700 pounds, and there were 1,300 pounds of lead inside, molded to fit the frames.

Generally uncomfortable, and entirely unfit for shallow water, the *Yolande* was not without her advantages. She was safer and had more accommodation than any other boat of the same water line, and could and did sail, in weather which sent the average centerboard craft scurrying for a sheltered harbor. In the ordinary summer weather, however, the centerboard of her length could sail around her with ease. Mr. Schuyler exhibited her weatherly qualities, by keeping her in commission until the snow began to fall, and showed that in bad weather, she could drown the centerboard boat completely.

She and the *Muriel* marked the introduction of a type of yacht that has undoubted advantages, but which, upon the whole, is not as well suited to the requirements of American yachting as is the centerboard, nor are they as a rule as speedy.

I may mention in passing, the building of another representative craft in July, 1879, and that was the iron centerboard sloop *Mischief*. She was the second sailing yacht built of iron in this country, and was a success as a racing vessel. We have had several iron yachts built since then, both sailing craft and steam, and I think that finally iron or mild steel will entirely supersede wood as building material for the pleasure fleet. Certainly it is best for steam yachts, and I think it better for sailing craft, as being lighter, dryer and stronger.

The cruise of the New York Yacht Club this year was marked by one of the old-fashioned regattas at New Bedford, for which, as I have shown, the club was famous in its early career. The entry was not, to be sure, a very famous one, but it made a fête day for the old whaling city, and will long be remembered.

There were two schooners, the *Tidal Wave* and *Phantom*, in the first class, and four, the *Magic*, *Peerless*, *Azalia* and *Clio*, in the second. There were also two classes of sloops, three in each. The winners were the schooners *Tidal Wave* and *Magic*, and the sloops *Niantic* (afterwards *Hildegard*) and *Vixen*.

October 17, 1879, there started four sloops from Sandy Hook Lightship, for a race around the Cape May Lightship and return, for a cup valued at $700, offered some years previous by Mr. Robert Center, then the owner of the iron sloop *Vindex*. He had successfully kept her in commission for a whole winter, defying the gale with the stoutest of pilot boats, but creating an impression in the minds of the hardy toilers of the sea in those boats, as they saw the *Vindex* under short canvas bobbing like a cork on the ocean swell, that "the gentleman was not just right aloft." They were unable to realize that any sane man should go to sea in such weather for pastime.

Mr. Center having demonstrated the ability of his iron keel vessel, cutter rigged, to withstand successfully all sorts of weather, determined with fine irony to show that the centerboard sloop could not do this; and so offered this cup for competition by sloop yachts in the month of October. For years the cup went begging, but in 1879 the *Mischief*, *Regina*, *Wave* and *Blanche* started for it; and this is not half as wonderful as that they all returned safely the next day. The "sweet little cherub" was certainly on watch during this race, for with the exception of the *Mischief*, four more unsuitable craft to be caught outside of Sandy Hook in October could not be found. I think it will be of interest if I give the dimensions of these "bowls" in which the four "wise men of Gotham" embarked:

NAME.	OWNER.	OVER ALL.		WATER LINE.		BEAM		DEPTH	
		ft.	in.	ft.	in.	ft.	in.	ft.	in.
Mischief	J. R. Busk	67	5	61	0	19	10	7	9
Regina	W. W. W. Stewart	59	8	47	3	19	3	5	0
Wave	Dr. J. G. Harrow	41	8	38	7	14	8	4	1
Blanche	C. H. Grundy	41	0	38	6	14	0	4	1

The *Mischief* was able to sail at least one-third faster than either of the others by reason of size, and as there was no time allowance she won with all ease. There was a moderate gale the day after the arrival of the yachts, and in some way a report got abroad that the *Wave* was missing, causing much uneasiness among the friends of those on board of her. As a remarkable race this is worthy of note here. I may mention also that the *Mischief's* time was 39h. 47m., beating the *Regina*, which came second, 4h. 20m. The *Wave* was third. In this connection, and having expressed an opinion as to the unsuitableness of shallow centerboard yachts of small size to encounter an ocean breeze and sea, I will give an illustration in opposition to that opinion. Early in the month of February, 1880, the sloop yacht *Coming*, having been purchased by a New York gentleman, he employed Captain Germaine and his brother of Glen Cove, L.I., to proceed to New London, where the yacht had wintered, and bring her to New York. Captain Germaine employed Mr. William H. Lane of New London to assist him and having bent the sails, they, as ordered by the owner, awaited a favorable chance to come to New York. It came in the shape of an offer from Captain Scott, of the tug boat *Alert*, who having been hired to tow the British brig *Guisborough* to New York from New London, offered Captain Germaine a free tow, and the *Coming* made fast to the stern of the brig and started. When a little to westward of New Haven, a hard northeast gale was encountered, and the tug finally, for her own safety, was obliged to let go the brig and make for New Haven for shelter. The brig made sail, but her sails were blown away and she finally sank off Northport, L. I., all on board perishing. Of the yacht nothing was heard for some days, when she was sighted off Southold, L. I., dismasted, with bowsprit gone, and port bow somewhat injured; but in all other respects in good condition. The anchors were on the bows, and the boats hung at the davits. In the cabin a meal of corned beef and cabbage was spread, and not a dish had fallen to the floor. The mast had fallen directly aft and lay on the deck, the wreck of the bowsprit and rigging was overboard, and this had operated as a drag keeping her head to the sea. Evidently the captain and crew, believing that they would be safer on the brig, hauled up under her counter to get on board of her, and in so doing the bowsprit and mast were carried away, and the bow stove. Had they remained on the yacht they would possibly have been saved.

This yacht, one of the extreme skimming-dish type, had safely weathered out one of the most terrific gales of that winter, and lived in a sea which was represented, by those out in it, to have been something tremendous. The life buoys and spare spars on her house were not lashed and were found undisturbed, showing that during her lonely drift not a sea had boarded her. This yacht was 61 feet, 4 inches over all; 56 feet, 10 inches water line; 20 feet, 5 inches beam; 5 feet, 2 inches deep, and 4 feet, 2 inches draught of water.

There is little to note of the yachting of 1880; the usual regattas and cruises taking place without any marked incident, except, perhaps, that this year another attempt was made at a handicap race by the New York Yacht Club; Mr. Charles Minton, the secretary, offering a $250 cup. The thing was a success so far as the handicap was concerned, and it is evidently the best of all systems for allowance; but the starters were few, only three schooners and six sloops. The schooner *Dauntless* and sloop *Mischief* were the winners.

I might also mention in passing that the first regatta of the Larchmont Yacht Club took place on July 5, 1880, its largest starter being the sloop *Viva*, 29 feet, 6 inches. As something of yachting importance I may say that the iron steam yachts *Corsair* and *Stranger* were launched at Philadelphia this year, the iron steam yacht *Polynia* was launched at Newbugh-on-the-Hudson, and the iron steam yacht *Yosemite* at Chester, Pa.—an evidence of the growing popularity of steam as a motive power among the yachtsmen, and this has been apparent more and more ever since and will continue. It may confidently be asserted that no more large sailing yachts will be built; but that all who can afford it will have steam.

During the cruise of the New York club this year, 1880, there was a fine race at New Bedford, the yachts of the Eastern and New Bedford clubs taking part, seven schooners and eleven sloops starting. The New York yachts *Crusader* and *Mischief* and *Regina* captured three of the prizes, and the New Bedford schooner *Peerless*—formerly a New York yacht—took the other. Yachtsmen in the fall of 1880 were a good deal fluttered by the rumor that the British cutter *Vanduara* was to come next season for the *America's* Cup. She was just then in the hey-day of her triumphs, and ranked as fastest in Great Britain, but has since been out-built and relegated to the second class. She did not come. Had she done so, I think she would have carried the cup back with her. We had very little respect for cutters in those days, and I presume would not have thought it worth while to have put anything better against her than the *Gracie*, *Mischief* or *Fanny*, in which case the *Vanduara*, on account of her extra size, would have had a sure thing. It was the golden opportunity missed that will not for a long time to come occur again.

The Eastern Yacht Club was in 1880 just ten years old, and it signalized the termination of its first decade by the purchase of a plot of ground on Marblehead Neck, and the erection thereon of a club house, which for many years was the finest yacht club house in the United States. It was a building seventy-five feet front, and three stories in height, furnished with all modern conveniences. It had on its roll in 1880, forty-three schooners, twenty-one sloops, four cutters and one yawl. Very

She was afterwards rigged as a cutter. She was 49 feet over all; 40 feet, 8 inches water line, 10 feet beam, 7 feet, 5 inches deep and 5 feet, 3 inches draught.

At a meeting of the Seawanhaka Yacht Club held November 20, 1880, Mr. M. Roosevelt Schuyler, then the vice-commodore, reported that he had been out sailing in his cutter *Yolande* two days previous, with three inches of snow on the deck. This was on the first introduction of the cutter, when its advocates thought it behooved them to show in all ways its superiority to all other types of boat. It probably never struck Vice-Commodore Schuyler

¹ "STRANGER."

many of the owners of the yachts, however, were more prominently identified with the New York than with the Eastern club, and the four "cutters" were such only in name, as neither in rig, or in shape of hull, did they resemble such boats as the *Bedouin*, *Wenonah* or *Muriel*. The yawl, however, was the *Edith*, and was modeled by Ratsey, of England, and built in 1880 by D. J. Lawlor at East Boston, and was the first of the rig built in this country.

that the owner of the shallowest of centerboards could have gone out in the bay sailing in a November snow storm if he had been silly enough to have desired to do so. Cutters were common enough after this, but I have not found that owners of them cared to keep them in commission any longer than it was comfortable to do so.

It was in March, 1881, that we again heard of a challenge for the *America's* Cup. It

¹ Cutter "Stranger." Owned by Mr. John N. McCauley, New Haven.

came again from Canada, and was prompted solely by the desire of Captain Cuthbert, the builder of the schooner *Countess of Dufferin*, for the advertisement and consequent increase of business which the notoriety of building a challenging yacht would give. The schooner he had built had proved a failure, but he asserted he could build a sloop which could beat any of the American single stick vessels, and a schooner could not be put against her with any chance of success, because there was, in the New York Yacht Club rules, no allowance for difference of rig.

The Royal Canadian Club had had enough of Captain Cuthbert, and of challenges for the *America's* Cup, but there was a spirited little club at Belleville, Ontario, with an attaché of the local newspaper as its secretary, and its members were delighted with the prospect of being brought prominently into notice as the challenger for this celebrated trophy; so probably for the first time outside of Belleville, Ontario, the Bay of Quinte Yacht Club was heard of. In the course of the preliminary correspondence that ensued, a writer in one of the New York weeklies incautiously suggested that the Bay of Quinte Yacht Club was hardly as important as the New York Yacht Club, and that the social position of the members of the latter was, perhaps, rather more elevated than that of the members of the challenging club; and he raised such a storm

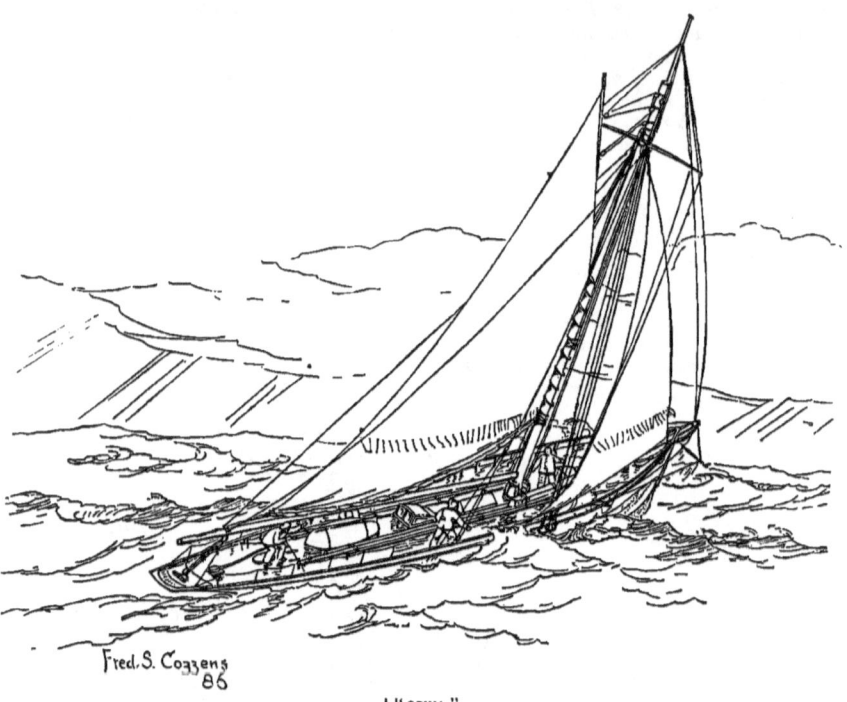

1 "ORIVA."

of indignation in Belleville that he repented his incautious utterance in sackcloth and ashes. However, the Bay of Quinte Yacht Club at its annual meeting adopted a resolution to challenge for the cup, and named September as the month for the race, or races.

At this time the flag officers of the New York Yacht Club were Com. John R. Waller; Vice-Com. James D. Smith and Rear-Com. Herman Oelrichs. These gentlemen had no doubt but that either of the sloops *Gracie*, *Mischief* or *Hildegard* would be fast enough to beat the new sloop building in Canada to compete for the cup; but with commendable spirit they resolved that if there was anything better in this country it ought to be at the disposition of the club. The

1 Cutter "Oriva." Owned by Mr. C. Lee Smith, New York.

sloop *Arrow* had, at that time, the best reputation for speed. She had been built in 1874 by Mr. David Kirby, of Rye, Westchester county, for Mr. Daniel Edgar, of the New York Yacht Club, and first appeared at the annual regatta of the club June 8, 1876, showing a wonderful turn of speed, and in ever saw the yacht until she was launched. They designedly refrained from all interference, and trusted to the builder of the *Arrow* to produce a sloop which should be, as he had promised, "swifter than the *Arrow*."

But the *Gracie* had been altered and much improved, and the *Mischief* had been built

[1] "ATLANTIC."

all subsequent matches she was easily fastest of the lot. She had been sold to Mr. Ross Winans, of Baltimore, who did not belong to the club, and who, in 1881, was abroad. The first idea of the flag officers was to telegraph Mr. Winans and offer to purchase the *Arrow*, but her builder came to them and said he could build a better boat than the *Arrow*, and they at once gave him *carte blanche* to do so. The result was unfortunate, but it was no fault of the gentlemen interested, neither of whom, I believe, since the time of the *Arrow's* triumphs, and both of these sloops were even then "swifter than the *Arrow*," and as was afterwards abundantly proven, much more speedy than the *Arrow's* successor from the shipyard at Rye.

May 26, the New York Yacht Club accepted the challenge of the Canadian club, assented to September as the time of the contest, thus waiving the six months' notice and all other formalities, as it always has done. The name of the challenging

[1] Sloop "Atlantic." Owned by Messrs. L. A. Fish, J. R. Maxwell and N. D. Lawton, New York. (One of the "big four" built to compete for the honor of representing America against the "Galatea" in 1886.)

sloop was the *Atalanta*, and according to the official certificate accompanying the challenge, she was about forty-five tons, and measured 70 feet, 1 inch over all ; 62 feet, 10 inches on the water line, 19 feet beam, 6 feet, 10 inches deep. She drew 5 feet, 6 inches aft, and 3 feet, 6 inches forward. In all respects, she was an American model, pure and simple.

The prospect of the international race gave an impetus to yachting this year as it has always done, and the regular annual events were more generally attended than for the few preceding years, and the contests more spirited. There was, however, nothing occurring at either of them that calls for special mention. It is interesting to note that the Larchmont Yacht Club at the time of its second annual regatta on the Fourth of July had enrolled thirty-six yachts.

After a pleasant correspondence, all the preliminaries for the race for the *America's* Cup, under the challenge of the Bay of Quinte Yacht Club, were amicably arranged ; the Canadian club naming the sloop *Atalanta*, and the American committee, Messrs. William Krebs, J. F. Tams and Robert Center, after consultation with the flag officers, assented to the request of the challenge, that only one yacht be named against the *Atalanta*.

As to which sloop this should be, there was considerable controversy. We had four fast vessels of about the required size, *viz.* : the *Gracie*, *Mischief*, *Fanny*, and *Hildegard*, and in addition to these, there was the new yacht building at Rye for the flag officers of the club, and to be called the *Pocahontas*. She was 71 feet, 6 inches on deck ; 65 feet water line ; 21 feet beam and 7 feet, 10 inches deep. Her centerboard is 21 feet. It is not necessary to give the dimensions of her spars, except to say that they were found to be too taut and had to be reduced. The *Pocahontas* was a failure. She had a fine entrance, but was too heavy in her counters for fast sailing. I have always thought that if lengthened aft and fined down at that end, she would make a fast schooner.

I note on August 16, 1881, the arrival of the steamer *Devonia ;* not a very remarkable circumstance considered alone, but the fact that she had upon her deck the little Scotch cutter *Madge*, made her arrival an important event in the history of American yachting ; for the result of the races sailed by her subsequently, did more to shake the faith of American yachtsmen in the superiority of the broad and shallow centerboard boat, than anything that had ever occurred.

It matters not whether her victories were won fairly or unfairly ; they were won, and the American sloop was for the first time defeated, and no excuse could palliate that.

This little craft was sent to this country by Mr. James Coates, the thread manufacturer of Paisley, Scotland. She was built at Gowan, Scotland, by Watson, in 1879, and was 46 feet, 1 inch over all ; 38 feet, 9 inches water line ; 7 feet, 9 inches beam ; 7 feet, 6 inches deep, and 8 feet draught. In Great Britain she rated as a ten tonner, but by the New York club rule she measured sixteen tons. Her skipper, Captain Duncan, with a crew of two men, came over with the yacht, and her subsequent success was largely due to the admirable and skillful manner in which she was handled.

I shrewdly suspect that the advent of the cutter *Madge*, and the races that were arranged for her after her arrival, were the result of the pre-arranged scheme on the part of some of the young gentlemen of the Seawanhaka club, who not only believed the British cutter to be superior to all other types of yacht, but were extremely impatient because everybody else did not think so. So they selected this little yacht, which had won many races in England, and then arranged some races for her under the Seawahaka rule of measurement, by which she was sure to win.

In justice to the *Madge*, I may say that she did not need the allowance at all, under the circumstances ; but the intention of these gentlemen was none the less worthy of remark. In furtherance of this scheme, three races were arranged for her with the sloops *Schemer* and *Wave*. I have never had the least doubt, but that either of these yachts, if in perfect racing condition, could have beaten the *Madge ;* but when the races were sailed, the season was near its end, the sails were fitting illy, and so little was thought of the chances of the *Madge* that not the least care was taken to put the American boats in racing order. At the first race, the American yacht had a borrowed topsail, which set "like a purser's shirt on a handspike," to use the forecastle expression, and the expert in charge of her said, when his attention was called to this, " Oh, it will do well enough, anything will beat that thing"; with a contemptuous gesture toward the *Madge*, which was lying at anchor with one of the most perfectly fitting suits of canvas I ever looked at. Previous to the races, whenever the *Madge* encountered one of the American sloops, the canny Scotchman in charge of her allowed her to be easily beaten, and it was had gone home, and the owner of the *Madge* was too shrewd to allow her to sail without him.

She has never done much since that time, but it cannot be denied that she "got her fine work in" very effectively during this first season, and established the cutter model in this country on a firm foundation, modifying and improving the American centerboard model, the result being a yacht like the *Puritan*, with the depth, the outside ballast, and in part, the rig of the cutter, retaining still the advantage of beam and centerboard.

The Canadian sloop *Atalanta* was launched at Belleville, Ontario, September 14, 1881, and by a curious coincidence, she and the *Pocahontas* had their first trial on the same day, October 5, the *Pocahontas* having a trial with the *Hildegard* and being beaten by her, and the *Atalanta* a spin with the *Norah* at Belleville, and beat her with ease.

The Canadian sloop could not be gotten ready in time for the race for the cup, and the request of the

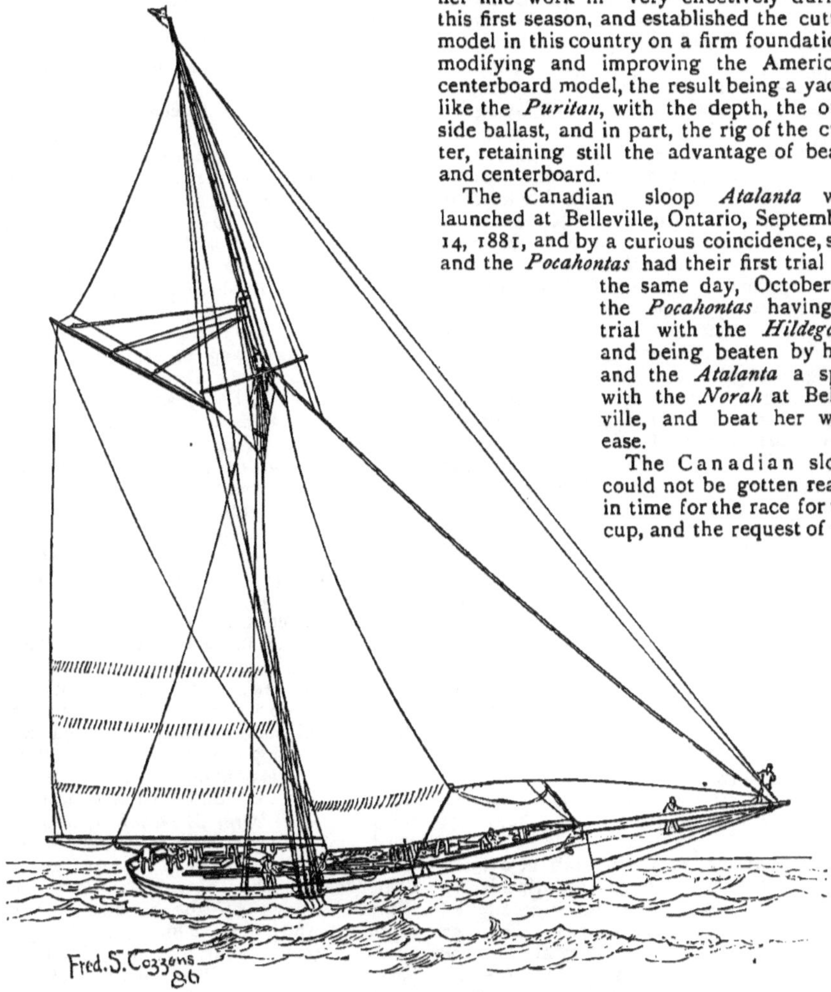

[Fred. S. Cozzens 86]

[1] "PRISCILLA."

not until the first match began, that any of us had ever seen her sail.

That she was a smart little craft is undeniable, and she was splendidly sailed. The owners of the *Schemer* and *Wave* went to much expense to fit their yachts for another race next season, but Captain Duncan Canadians for an extension of time was cheerfully granted by the New York club, and meanwhile a series of trial races was arranged, the entries for which were the *Gracie*, *Mischief*, *Hildegard* and *Pocahontas*, but the *Hildegard* withdrew after one trial. The choice very soon narrowed down to the

[1] Sloop "Priscilla." Owned by Mr. A. Cass Canfield, New York. (One of the "big four.")

Gracie and the *Mischief*, and the latter was finally chosen.

The Canadian sloop finally arrived, *via* the canals, October 30, and the two races with the *Mischief* were sailed November 9 over the course of the club, the Canadian being beaten 28m. 30¼s., and November 10, over a course outside the Hook, the *Mischief* again winning by 28m. 54s.

The irrepressible Captain Cuthbert at once announced his intention of laying his sloop up in this harbor, and renewing the challenge the next season, and to protect itself against this threatened annual Canadian infliction, the New York Yacht Club was obliged to insist upon such a change in the deed of gift of the *America's* Cup as would prevent this. It therefore returned the cup to Mr. George L. Schuyler, the only surviving donor of it, and received it back from that gentleman with a clause providing that a defeated yacht should not be again eligible as a challenger until two years had intervened from the time of the first contest. At the first meeting of the New York Yacht Club in 1882, a proposition was made and afterwards adopted, to do away with the club uniform, a decided improvement, and at this meeting also, Mr. Ogden Goelet, the owner of the fine keel schooner *Norseman*, advised the club of his intention to present two cups, one of $1,000 for schooners, and one of $500 for sloops, to be raced for off Newport during the annual cruise of the club. Mr. Goelet repeated his liberal donation each year for some years, and the Goelet Cup race became finally the most important event of the yachting season. Newport being half-way 'twixt Boston and New York, the race for these cups was always participated in by more or less Eastern yachts, the famous sloop *Puritan* scoring here her first victory.

The Seawanhaka Yacht Club, at a meeting held March 2, tacked on Corinthian to its beautiful Indian name, and was weighted down with it for several years. The idea was, as stated by the advocate of the change, that this club, having been the first to introduce Corinthian yachting, ought to have something in its name to call attention to the fact; that so many clubs were now adopting the Corinthian system, the glory of its introduction would be lost to the Seawanhakas if they did not in some way label themselves as "the only true and original Jacobs." It was a snobbish reason for an ugly suffix, and it weighted the club down terribly, at one time nearly carrying it under entirely. I may mention also that the Seawanhaka club about this time changed its rule of measurement, adopting the "sail area and length" rule, which, although not as favorable to the cutter as the old rule, was still very much in favor of this type of yacht.

It was in 1882 that the British cutter *Maggie* was imported, having been brought over as the *Madge* was, on the deck of a steamer. She was a fifteen tonner, and of her *Bell's Life* said: "We are free to confess that she is the best fifteen tonner which has ever carried a racing flag in this country." The *Maggie*, however, has not done much here, having been repeatedly beaten by centerboard sloops. In fact, there has never been a square race between the cutter and the sloop, but what the sloop was proved the victor. In extremely light weather the cutter has generally been able to win, but in strong breeze with smooth water the sloop has always come off conqueror. It was in this year that the cutters *Bedouin* and *Wenonah* were built at Brooklyn by Henry Piepgras, and taking all things into consideration, the *Bedouin* has been a most successful yacht.

The usual regattas and cruises of the clubs took place this year, but there was nothing in connection with them at all noteworthy except that the New York Yacht Club on its cruise went around Cape Cod, and sailed a race at Marblehead; and at its close the centerboard sloop *Vixen* had a match with Mr. Warren's imported *Maggie*, and beat her very decidedly.

As an appropriate wind-up to the season, the Seawanhaka Corinthian club organized a series of sloop and cutter races, making two of the series outside the Hook, and in the full belief that under such conditions the cutters *Bedouin*, *Wenonah*, and *Oriva* must win. They were much disappointed at the result, having been in favor of the centerboards, the *Gracie*, *Valkyr* and *Fanita* carrying off the honors. I note April 7, 1883, the launch of Mr. Jay Gould's steam yacht *Atalanta*, from the yard of the Messrs. Cramp, at Philadelphia, by all odds the finest yacht ever built in this country.

At the May meeting of the Eastern Yacht Club, Mr. Jay Gould, the owner of the *Atalanta*, was proposed for membership and rejected, and there is every reason to believe that the only reason his name was not proposed in the New York Yacht Club was, that it was quite certain that if proposed he would be rejected there also.

This reminds me very much of the little girl of the story book, who refused to eat her breakfast, just to spite her mother. Any yacht club ought to have been proud to have enrolled so splendid a yacht as the *Atalanta*, or for the matter of that, a man as influential as Mr. Jay Gould. From this action of the Eastern club and the probable action of the New York Yacht Club, resulted the organization of the American Yacht Club, to consist principally of owners of steam yachts, and to which, in time, all owners of steam yachts must inevitably be attracted. I think that whoever in the year 1900 shall continue the history of American yachting, will speak of the American, as the principal yachting organization of the United States.

There were two schooner yachts launched in the early part of the year 1883, which became very prominent afterwards. The one was the keel schooner *Fortuna*, built by the Poillons at Brooklyn, for Commodore Henry S. Hovey of the Eastern club, from a design by A. Cary Smith, and the other the centerboard schooner *Grayling*, built by the same firm for Mr. Latham A. Fish of the Atlantic club, from a design by Mr. Philip Ellsworth. Soon after going into commission, the *Grayling* was struck by a squall while sailing in the lower bay, and capsized and sank. She was raised and refitted; the principal result of the accident being to bring into prominence the indomitable pluck and perseverance of her owner, who in eighteen days from the time she sank, had her ready to start in the Decoration Day's sail of the club.

In the earlier days of yachting in this country, as I have shown, the sloop *Julia* figured as fastest in the fleet. She had been sold to an eastern man and rigged as a schooner. In the early part of 1883 Mr. Edward M. Brown, then Rear-Commodore of the New York Yacht Club, purchased the *Julia* and had her rigged as near as possible as she was in the time of her early triumphs; many of the older yachtsmen believing that no improvement in model had been made in the quarter of a century that passed since the *Julia* was built, and that the old yacht in her old form would beat any and all of the modern productions. They were mistaken, just as the people are nowadays who think that the old *America* is as fast as the modern schooner. The fact is, that we have constantly improved both in model and in rig.

It was also in the early part of this year, 1883, that the cutter *Marjorie*, since so celebrated, was launched at Greenock, and it was rumored that she was to come here for the *America's* Cup. In the light of subsequent history, I think that there is good reason for saying that if she had then come, she would have carried it home with her. We had not much opinion of the speed of cutters at that time, and I don't think, after the experience of the *Pocahontas*, that anything would have been provided to sail against the *Marjorie* except either the *Mischief*, *Gracie*, or *Fanny*.

The clubs, as usual, had their annual regattas, only notable from the fact that this year, the New York Yacht Club once more changed its system of measurement for time allowance from the cubical contents rule to that of sail area and length. It was not that the old rule had not proved satisfactory, for it had; but it was felt to be desirable to adopt some rule more favorable to the cutter, so that this style of boat could be induced to enter in the sloop class, and to prevent the necessity of having a special class for them. There were now the *Bedouin*, the *Wenonah*, the *Oriva*, the *Muriel*, and others in the club, and there was desire on the part of the club members to give them a chance. The rule is acknowledged to be an unfair one for the sloop, and I presume would have been changed, but for the fact that under it, two challenges from cutters have been accepted, and it could not consistently be changed until these races were sailed.

The other notable event in connection with the annual race, was the sailing of the Atlantic Yacht Club regatta in a thick fog, and the colliding of the Committee steamer with one of the Norfolk steamers as she was returning from the lightship. Fortunately no one was injured on either steamer, although both vessels were much damaged.

The fleet of the New York Yacht Club, on its annual cruise this year, went by invitation of the Eastern Yacht Club to Marblehead, and sailed a race there. Mr. James D. Smith was the commodore at this time, and so popular was he, that he carried a fleet of fourteen schooners and ten sloops around Cape Cod. The club had also the tug *Luckenbach* under charter, and had her accompany the yachts in order to render prompt assistance, should it from any cause be required.

The regatta at Marblehead was sailed August 10, and the number of starters was not very large. It included but four first-class and five second-class schooners, and four first-class and three second-class sloops. The cutter *Wenonah* at this race beat the sloop *Mischief* over a minute, a pretty correct indication of what would have happened had the *Marjorie* come that year for the *America's* Cup.

As showing the progress of yachting in this country, I may mention the fact that on August 18, the Beverly Yacht Club had a regatta at Marblehead in which there actually started 171 yachts. The largest was the cutter *Wenonah*, of 66 feet mean length, and the smallest the cat-boat *Faith*, 14 feet, 8 inches. October 16, of this year, the Seawanhaka Yacht Club had a fall race, the first of a series of three contests that it had arranged for sloops and cutters. Only the cutter *Bedouin* and the sloop *Gracie* started, the wind was strong and the sea heavy, and of course the cutter won as she liked.

Early in the season, a match had been made between the sloop *Gracie* and the cutter *Bedouin* to race in October, outside the Hook, for $1,000, and this race was sailed October 18. There was not much wind, but there was a heavy roll, as a result of the strong breeze of the previous day. The cutter beat the *Gracie* 15m. 5s. on corrected time. The sloop, however, had her innings two days later in a race outside with a smooth sea and a strong lower-sail breeze, when she beat the cutter with ease. This was quite a season for match races, and on October 25, the sloop *Fanny* defeated the *Gracie* in a match for $1,000, outside the Hook. Neither yacht was suited to ocean racing, but the wind was moderate and sea smooth, so both came off without accident, and this closed the racing of the season of 1883.

I find nothing of note in 1884 until June 14, when the Seawanhaka club had its annual Corinthian regatta in a moderate gale, and of its eight starters, only three — the *Gracie*, *Oriva* and *Petrel* — finished, the result showing conclusively that in heavy weather the centerboard yacht has no business outside of Sandy Hook. Just then, the cutter and sloop controversy was raging fiercely, and the result of this match made the cutter advocates jubilant.

Yachting in the New England States continued to increase more rapidly than in any other section, and a muster roll of the Boston Yacht Club for this year shows twenty-four schooners, thirty-two cabin sloops, ten cat-rigged boats, six cutters, eleven steamers and a catamaran. Most of these were distinctively Boston club boats, and did not, as was the case notably with the Eastern club, owe prime allegiance to another organization. Among the steamers was Jay Gould's *Atalanta*, and among the schooners, the old *America*.

There was a race around Long Island during the season of 1884, but, as has been the case with all races over long courses, the result was unsatisfactory. The element of chance enters too largely into the result. In this case, although the *Grayling*, undeniably the fastest schooner, won, her victory was due to good luck and skilful handling during the last twelve hours of the contest. There were six schooners, five sloops and three cutters. The cutters were badly beaten, and sloop stock was once more buoyant.

In July of 1884, Mr. William Astor's steam yacht *Nourmahal* was completed at the yard of the Harlan & Hollingsworth company, after nearly a year spent in her construction. She is 250 feet long, and, the finest yacht in the country, except, perhaps, Mr. Gould's *Atalanta*.

In August, 1884, the American Yacht Club had its first steam yacht race, over the course from Larchmont to the entrance of New London harbor, a distance of about ninety-two miles. Of course the arrangements were far from perfect, the thing being almost in the nature of an experiment; but it was proven that races of steam yachts could be satisfactorily arranged, and with better results the race has been repeated each year since that time.

The Seawanhaka club had its usual fall match for sloops and cutters this year on October 18, and for the cutter advocates it proved very successful. Out of a lot of fourteen starters, not a sloop showed up at the finish line. The only ones which finished were five cutters. The race was sailed in a howling nor'wester, and the sloops could not stand the press.

In December of 1884, we learned that the owners of the cutters *Genesta* and *Galatea* were about to challenge for the *America's* Cup, and immediately all was excitement, not only among yachting men, but among the general public. In fact, I think there was more interest taken in the affair by persons outside of the New York club than by its members.

One and all recognized that these were challenges from very different yachts from the *Countess of Dufferin* or *Atalanta*. We had come gradually to have much more respect for the cutter model than at first. The *Bedouin* had shown herself quite as good as the *Gracie*, the *Oriva* had proved herself better than the *Vixen*. The record of the *Genesta* was familiar to all American yachtsmen, and the new yacht building was presumably better than the *Genesta*. So with wonderful unanimity yachting men agreed that if the cup was retained it must be by a yacht yet to be built, for neither of our four fastest sloops could hope to retain it.

Mr. James Gordon Bennett was the commodore, and Mr. W. P. Douglass the vice-commodore of the New York club, and they at once resolved to build a yacht about the size of the *Genesta*, and after careful consideration they accepted the design of Mr. A. Cary Smith for an iron sloop, and gave the Harlan & Hollingsworth company the contract to build her. Commodore Bennett at one time resolved that he would have in addition a wooden yacht from a design by Capt. Philip Ellsworth, but finally relinquished this and concluded to trust the defense of the cup to the *Priscilla*, as the new yacht was to be named.

Meantime plans for yachts to defend the cup poured into the New York Yacht Club rooms at the rate of one or two a day, and we never before fully realized how much of architectural talent we had. Many of these plans were meritorious, and many more bore the impress of the brains of "cranks."

Meanwhile, several gentlemen, members of the Eastern Yacht Club and also of the New York; men of great practical experience in yachting, and also men of more than ordinary intelligence, had pondered and agreed upon a design for a centerboard yacht that should combine all the advantages of the cutter's model and rig, with the best features of the American model and rig. The result of this combination of brain and practical experience, is the sloop yacht *Puritan*.

Her design is credited to Mr. Edward Burgess, of Boston, but I consider him as but one of four to whom the credit should be given. The *Puritan* has been called "a happy accident," but in point of fact there was nothing accidental about her. From stem to stern, from keel to truck, all things about her were closely calculated. She has the keel and outside lead of the cutter, and the centerboard of the sloop. She has the short mast and long topmast of the cutter, the straight round bowsprit of the cutter (and if she could have had it fitted to house as the cutters do it would have been an improvement) and on this her jib sets flying, as in the cutter. Her mainsail is laced to the boom as in the sloop, and in this respect the cutter people are copying her fashion.

This yacht was of wood, and was built by G. Lawley & Sons, at Boston, and proved superior to any yacht ever built in this country, not only for speed, but for sea-going qualities. She proved herself able to beat the *Genesta* in ordinary racing weather, and in real bad weather, I have no doubt, her superiority would be still more apparent.

The yacht built for the flag officers of the New York Yacht Club proved also extremely fast. But for the advent of the *Puritan* she would have been considered a marvel. Tried with the *Puritan*, however, in a race off Newport for the Goelet Cup, in very ugly weather, the superiority of the Boston sloop was so plainly apparent, that it was evident to all that she must be the chosen yacht. Some changes were made in the *Priscilla*, and a series of trial races was sailed here, the result being the choice of the *Puritan* to sail against the *Genesta*.

I may not dwell on the details of those races, and it is not necessary, for they must be fresh in the minds of most of my readers. The series of races arranged, consisted of one contest over the course of the New York Yacht Club, one twenty miles to windward and return outside the Hook, and one over a forty-mile triangle outside. As was the case with the two previous cup contests, only two races were necessary; one over the inside course sailed September 14, 1885, resulting in a victory for the *Puritan* of 16m. 19s. corrected time; and one sailed September 16, over a course twenty miles east-southeast from the Scotland Lightship and return, won by the *Puritan* by 1m. 38s. corrected time. The races demonstrated that the *Genesta* was an exceptionally fast vessel and could probably have beaten any other sloop in the country save the *Puritan*.

September 18, she started again in a race for a $1,000 cup offered by Vice-Commodore Douglass, over a forty-mile triangle outside, and she beat the *Gracie* 21m. 52s. Races had been arranged for the Brenton's Reef and Cape May challenge cups, and for these the only yacht which started against the *Genesta* was the schooner *Dauntless*. The result was a foregone conclusion from the start, and in fact the intent of the club members was to allow the *Genesta* to take these cups to England: First, because they had proved nuisances here, and second, because they wished to have something to go to England for, if any owner should so desire. This finished the career of the *Genesta* in this country, and she left for England, October 8.

This also closes the yachting for 1885, and with this I will end this history of American yachting. I should have been glad to have made it more full and complete, but have been obliged to omit mention of all except the most important events. I have intended to make it as much as possible a record, as well as to show the well-nigh marvelous growth of the sport in the short space of forty-one years.

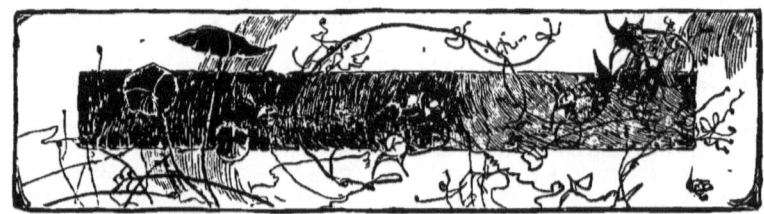

THE MAYFLOWER AND GALATEA RACES OF 1886.

THE MAYFLOWER AND GALATEA RACES OF 1886.

BY CHARLES E. CLAY,

Author of "BERMUDA YACHTS AND DINGHIES," ETC.

ENGLISH yachtsmen have made another effort for the recovery of the *America's* Cup, a trophy that has come to be regarded as the emblem of the supremacy of the seas, and that effort has met with a defeat more disastrous and humiliating than that which attended the unsuccessful attempt of the *Genesta* last year.

Scarcely had Sir Richard Sutton berthed his favorite in her snug winter quarters than Lieutenant Henn challenged for the ensuing year. In this he was more patriotic than wise, for, while nobody denies that the *Galatea* is a thoroughly representative type of the highest development and perfection of the English model, yet it cannot be conceded that her performances were enough, if any, superior to those of the *Genesta* to warrant her owner having any valid grounds for supposing his boat could do what her sister had failed to achieve.

If Lieutenant Henn felt enthusiastic enough to enter into a competition that for the past thirty-six years has baffled the highest naval architectural talent of Great Britain, would it not have been more prudent to have set to work during the winter and built a yacht more after the type and model of the one that had vanquished the *Genesta*, built by the same designer, and embodying every principle contained in his own boat? Surely Mr. Beavor Webb is not so hopelessly wedded to his own designs and ideas as not to perceive and appreciate the good points and qualities in the productions of a rival, and a successful one at that. If the results of the last two years' contests point to any conclusion at all, it is that the decided success of the American boat is not due, one iota, to the favorable condition of wind and wave, as is the universal howl of the rabid cutter men, but is inherent in the superiority of the principles involved in the construction of the model, and I contend most emphatically that so long as English yachtsmen go on building a V-shaped, leaded plank-on-end type of boat, simply because "they are so much better adapted for our waters," without ever giving the American type a fair trial, just so long will America continue to hold the yachting "blue ribbon." It is not enough for Englishmen to send one boat after another of the same type, just because each successive aspirant is claimed to be better than her predecessor. Change and modify the model from the bitter lessons that have been taught us, and then, and not till then, may we hope to compete with some reasonable prospects of victory.

The general supposition among us in England to-day is that, given a gale of wind and a heavy, choppy sea, and there is nothing like a deep-keeled cutter with an enormous weight of lead attached, to thrash to windward. This may be undoubtedly the case with regard to the types of boats with which the majority are familiar; but it does not apply to the newest type of the American centerboard sloop, a type not known in British waters, nor to English yachtsmen; and recent trials and the most thorough tests go to prove that the *Mayflower* in a sea way is superior in many of the most essential qualities of a rough-weather craft; she does not bury and "*hang*" so long when pitching as the English model; she has a quicker recovery and rides *over* and not *through* the sea; she points up as high, and eats her way as well to windward, besides being faster.

But I am afraid I have digressed somewhat from the object of this paper, which is to give a description of the actual incidents of the all-absorbing races, rather than a dissertation on the types and merits of the contestants.

No sooner was the challenge received than the leading clubs of the country set about seeing that nothing was left undone to retain a prize they had so long and so successfully owned. They might very naturally have said: "Well, the *Galatea* is no better than the *Genesta;* and the *Puritan* can do for the new-comer what she did for her sister." But that is not the spirit of the American people; they never rest content with what they have; the future is always sure to produce a better article than the best of the present. This noble spirit of emulation brought four competitors into the home lists; of them two were old favorites; the *Puritan*, trusty, stanch, and bearing the laurels still fresh upon her victorious prow; the *Priscilla*, with every defect altered, but still a novice eager to gain her maiden honors; the remaining débutantes were the latest skill of the builder's art; the *Atlantic*, which, however, never fulfilled the anticipations of her designer, and the queenly *Mayflower*, the fairest sea anemone that ever bloomed on American waters. All honor, then, to Boston, her birthplace, and to Mr. Burgess, her skilful designer.

The trial races were most satisfactory, and proved beyond a doubt that the *Mayflower* was the queen of the "big four," fairness could dictate, was handsomely made.

The first of the three courses to be sailed over (if three trials became necessary), was the one known as the regular New York club course, which, starting from a line off Owl's Head in the inner bay, leads out through the Narrows, rounding buoy 8½ on the port hand, and then on and around Sandy Hook lightship, and home again round buoy 8½, finishing off the Staten Island shore over a line somewhat to the northward of Fort Wadsworth. This makes a splendid all-round course of thirty-eight miles, and is eminently calculated to try the various points of sailing

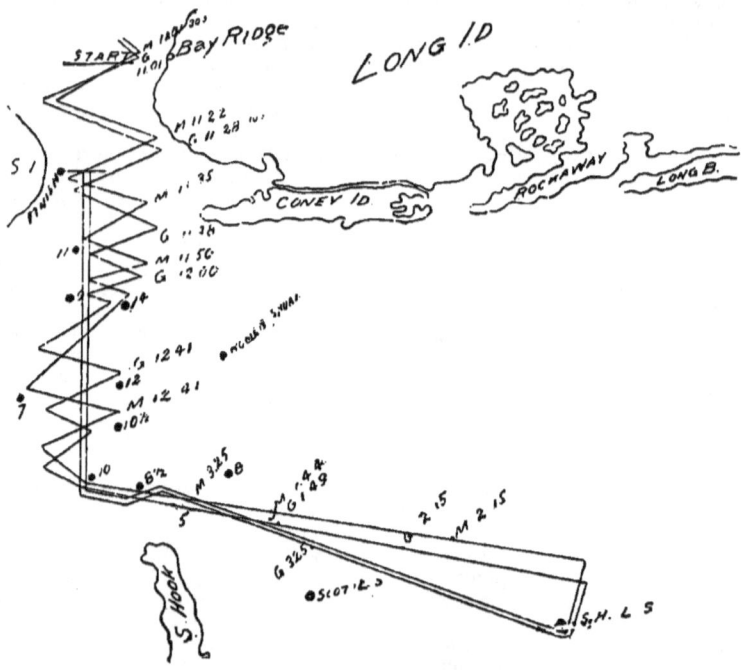

[1] THE COURSE.

and to her shapely hull and tapering spars might be entrusted the glorious distinction of doing battle for her country, let come what might. To the New York Yacht Club, the oldest and leading yachting organization in this country, was entrusted the honor of making the arrangements necessary to bring the impending struggle to a fair and impartial issue, and well did they perform their task. The gallant visitor was consulted on every point, and every concession that courtesy and on, off, and before the wind. Over this course the sloop is supposed to have a slight advantage, as comparatively smooth water and light winds are generally the rule on these waters.

The second and "outside course," as it is called, is a twenty-mile thrash to windward from the Sandy Hook or Scotland Lightship, according to the direction of the wind, with a run back. These conditions are favorable to the cutter, and chosen to make things square. Should each boat win

[1] From Captain Coffin's account in the *New York World*.

THE MAYFLOWER AND GALATEA RACES OF 1886.

one of the first two races, the deciding course would be a triangular one, but as it was not needed this year, the bearings need not be given.

THE FIRST RACE, TUESDAY, SEPTEMBER 7.

The all eventful *expectata dies*, so eagerly longed for by enthusiastic thousands, dawned with anything but a promise of fine weather or favoring gales. A dull leaden curtain hung over the busy city. Flags drooped limp and motionless against their poles, and with a heart full of misgiving I awaited the arrival of the *Stranger* at the Twenty-third street pier. Off in the stream lay the steam yacht *Electra*, while darting backwards and forwards her saucy little launch conveyed on board the guests of her owner, Elbridge T. Gerry, the commodore of the New York Yacht Club.

Soon, down the river from Mr. Jaffray's country place on the Hudson, came the *Stranger*, not only one of the handsomest and largest steam yachts in the world, but certainly the fastest of its size. Our courtly host lost no time in welcoming us on board the launch; we were speedily puffed out to the larger craft, and in a few minutes more good Captain Dand was heading the *Stranger* full steam down stream,

"To join the glad throng that went laughing along."

We did not lack for company; every conceivable craft was bound our way, from the leviathan excursion steamers with decks massed black with people, to the tiny skiff piloted by its solitary occupant. And now we are amid the flower of America's floating palaces, and close beside us steams the *Atalanta*. Beyond is the *Corsair*, with Lord Brassey aboard. Ahead, astern, and on every side are seen the gleaming hulls of beautiful yachts, the *Oriva*, *Orienta*, *Tillie*, *Puzzle*, *Radha*, *Magnolia*, *Vision*, *Speedwell*, *Ocean Gem*, *Theresa*, *Oneida*, better known as the *Utowana*, *Viking*, *Wanda*, *Nooya*, *Falcon*, *Electra*, *Vedette*,

"Cum multis aliis quæ nunc præscribere longum est."

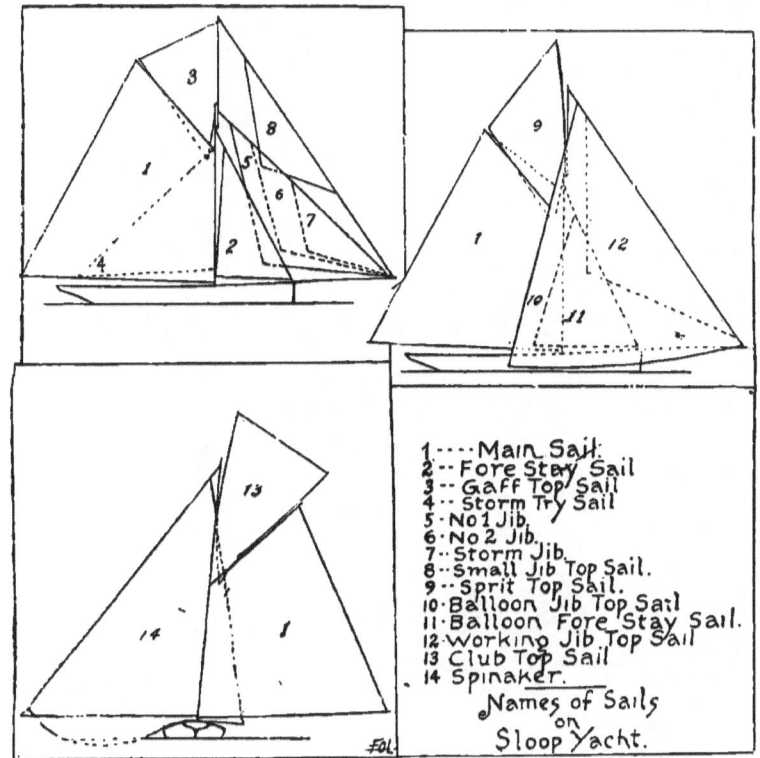

[1] DIAGRAM OF YACHTS.

[1] From Captain Coffin's account in the *New York World*.

The flyers of other days, too, are there, the *Rambler, Columbia, Ambassadress, Tidal Wave, Montauk, Ruth, Priscilla, Dauntless, Carlotta, Fleetwing, Mischief, Republic, Wanderer, Wave Crest, Gaviota* and a host of their fair sisters, whose names I could not get. And darting hither and thither among the fleet like some hissing, fiery snake, emitting from time to time the shrillest of piercing whistles, rushed the rakish-looking little steam launch *Henrietta*, Mr. Herreshoff's last production, said to go an average speed of twenty knots an hour.

Anxiously we scanned the distant Narrows to see if there was any sign of a coming breeze, and as if in answer to the silent ejaculations of the assembled multitude, a dark ripple was seen to ruffle the glassy surface of the bay, and gave promise of a breeze outside.

It was now ten o'clock, and the rivals were daintily picking their way in and out among the waiting armada, manœuvring to get a good start as the whistle bade them cross the line.

At the warning scream the *Mayflower* stood bravely for the line, carrying her boom to port with club-topsail, staysail and jib set, and breaking out her jib-topsail as she crossed. O! it was a beautiful sight, and made every pulse beat quicker, and sent the warm blood tingling through my veins. The British cutter was not a whit behind; "hauling to" very sharply, she rushed, with great headway, in between the sloop and the stake boat, and got the weather gauge, blanketing her antagonist, who had to keep off a trifle in consequence.

This was a very smart and seamanlike manœuvre, but in my humble opinion it was an error in judgment, for, had the cutter taken the leeward place, with her pace at the time, she could have stood the detriment of the blanketing for the short time they held the starboard tack, and when she went about, would have compelled the sloop to do the same, and so had the *Mayflower* under her lee for the long leg over to Staten Island.

However, the fact remains that, despite the *Galatea's* blanketing, the Boston sloop ran away from under the Englishman's lee, and when the latter, owing to her deeper draught, went about off Bay Ridge, the *Mayflower* stood on for another thirty seconds and came about well to windward, and had the cutter where she wanted her, and where she kept her till she was a beaten boat.

Off Fort Wadsworth, the two boats again tacked, the *Mayflower* at 11:13:30, and the the cutter a minute later, and stood across

"GALATEA."

to Fort Hamilton. Two things now quickly became apparent: that the *Mayflower*, though sailed a good rap full all the time, pointed just as high as the *Galatea*, which was evidently being sailed very fine, as shown by the continual lifting and shivering of her head sails, and, that the saucy Yankee had the heels of her English rival and was creeping ahead and to windward very fast. At 11.22, the *Mayflower* and showed a want of courtesy that no real "salt" would have thought of being guilty of. At 11.35 the *Mayflower* went about off Gravesend Bay, and the *Galatea* followed suit at the same moment, a little to the southeast of buoy No. 15. In these repeated tackings, it was noticeable that the *Galatea* was the handier "in stays," the American craft appearing just a trifle sluggish. On entering the Narrows the breeze seems to be freshening up a little, and the Yankee boat bends gracefully over to it, and the white spray dancing round her bows shows that she is quickening her pace. The *Galatea* stands up straighter, and is slipping through the water without much fuss, but does not seem to be gaining much on her fleet-winged rival. Off buoy 13 the *Mayflower* went "in stays" again at 11.41½ and stood towards

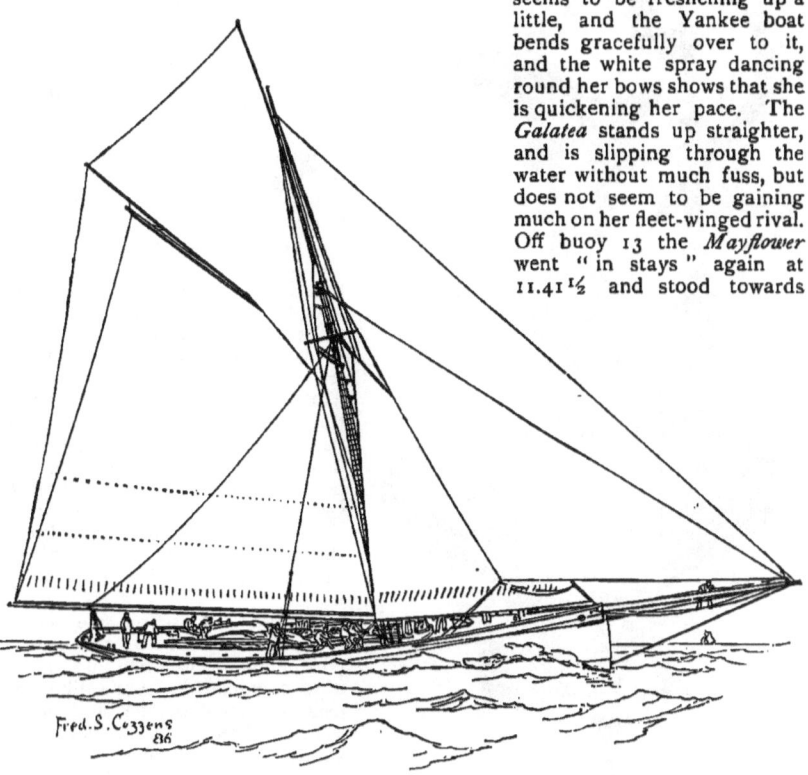

"MAYFLOWER."

went about again, and stood on a long reach into the Narrows to get the benefit of the slackwater. Ten minutes later the *Galatea* tacked and stood towards the Staten Island shore, but the *Mayflower* had gone about again and stood towards the Englishman, whom she cut about 200 yards dead to windward. While the *Galatea* was on this tack, the *St. John*, the regular Long Branch steamer, had the bad taste to sail right across the *Galatea's* bow, treating her to all her backwater. It was a churlish act, Coney Island Point, and six minutes later she was followed by the cutter. The sloop made but a short leg here, and at 11.50 she went about again, bringing both boats on the same tack, heading about east. The sloop seems to have doubled her vantage of 200 yards. They seem to be sailing the cutter a bit fuller now, but as we pass astern of her I notice that she has her weather jib-topsail sheet towing in the water On this board the cutter appears to gain slightly on the sloop, and at half a minute before noon she goes about once more; the *Mayflower* follows her lead at 12.03½, and goes round between buoys 9 and 11. The recital of tacks seems endless, but on each board the American boat increased her lead, and finally rounded buoy 8½ at 1:1:51, official time. The *Galatea* weathered the same buoy at 1:7:7. From

THE "AMERICA'S" CUP.

here to buoy 5 the positions of the contestants did not vary much, and the *Mayflower* led her antagonist by about six minutes, irrespective of the 38 seconds time allowance she had to give the cutter. The wind continues light, and the sea is as smooth as a tennis court. Rounding buoy 8½ both boats can about lie the course to the lightship, which bears S.E. by E. The breeze seems a good deal fresher outside, and the *Mayflower* is dancing gaily along, lying over to her plank-shear. How gloriously buoyant is her motion as she rises and falls to the gentle undulations which make up as we gain the open water! This is the longest reach of the day, and gives us all a breathing spell for refreshments.

At 2.28 the sloop comes "in stays," and takes in her jib-topsail as she stands towards the ugly-looking red hulk that shows the way into the channel. Her crew are busy getting her balloon jib-topsail run up "in stops," and soon a white streak running from truck to bowsprit end appears. The floating navy that has accompanied us all the way are gathered thickly round the lightship, hovering like bees about a sugar barrel; and now, as the swiftly gliding sloop approaches the turning-point, their pent-up enthusiasm can be restrained no longer, first one and then another impatient tug and steamer emits her shrill scream of welcome, and then all at once it seems as if every demon from the nether world is let loose, roaring round the *Mayflower*. The toot-toot-tooting is simply ear-splitting. Cannon thunder forth their approbation from brazen throats; frantic crowds bellow themselves hoarse; the very planks beneath my feet seem starting from the seams of the *Stranger* as her booming cannon, withheld by rigid discipline till the exact moment of rounding, belches forth her quota to the hurly-burly around us.

But see! it is scarce five seconds since the *Mayflower* turned her sharp prow to plow homewards, when lo! a white puff of snowy canvas bursts like the smoke from a distant battery, and bellying to a spanking breeze, her balloon jib-topsail is sheeted home and envelops her from topmast head to end of her jibboom, and away aft to her full waist. Well and smartly handled, ye motley crew; you may not look so neat and natty as the uniformed lads of the *Galatea*, but the old Norse blood of your forefathers runs in your veins, and ye are no degenerate sons of Hengist and Horsa, and the other vikings of your native land.

But *Væ victis!* Already the tardy cutter is almost forgotten as she struggles bravely on, irrevocably handicapped beyond redemption now, for the sloop is running while she has still a weary beat before she can do the same. At last she too tacks for the turning mark, but carries her baby jib-topsail to the very last minute, in the hope of gaining a yard or two thereby. She tacked at 2.40, and at 2.44 is fairly off after her rival. Now, boys, bear a hand; up with your balloon; you have not a moment to lose; the breeze that favored the Yankee is fast dying away, and you must make the most of it. Why, what's the matter, ye hardy sons of Yarmouth? Ah, there it goes up!—up! What! it's surely not foul? Yes! down, down, it has to come, and three weary minutes are consumed before it gets to the topmast head, and begins to draw. The game is well-nigh over now; away in the distance, like some huge albatross with outspread pinions, the *Mayflower* is nearing buoy 8½, which she rounds at 3.34, and so round S.W. spit buoy 3½ minutes later, and jibed her mainsail to get her spinnaker under way. But the wind had hauled into the eastward, and the boom was left in slings ready to be dropped at a moment's notice.

The *Galatea* rounds buoy 8½ at 3.46¾, and the S. W. spit buoy at 3.50. The wind freshens a trifle, and the cutter tries her spinnaker, and the *Mayflower* follows suit almost immediately. The goal is rapidly neared now; the same demoniac noises commence, but are kept up twice as long, and, if it were possible, are twice as loud. The very bosom of the mighty deep seems to tremble, and, amid salvos of cannon, the jubilee of 50,000 throats, and the ovation, congratulations and rejoicings of such a multitude as had never before gathered on New York's historic bay, the peerless Boston sloop *Mayflower* bore her happy owner, General Paine, over the line at 4.22½.

I append the official time:

	START	FINISH	ELAPSED TIME	CORRECTED TIME
	H. M. S.	H. M. S.	H. M. S.	H. M. S.
Mayflower....	10 56 12	4 22 53	5 26 41	5 26 41
Galatea.....	10 56 11	4 35 32	5 39 21	5 38 43

Mayflower wins by 12m. 2s.

The conclusions to be drawn and the lessons taught by this momentous struggle were briefly these: that in light breezes and a smooth sea the English model, as represented by the *Genesta, Galatea* and *Irex* type, cannot compete at beating, reaching or running with the American build. That with regard to seamanship and expert handling of their craft the Americans have nothing to learn from their cousins from over the water. That, having at the outset been the humble disciples of the mother country, they have reached that stage in the science and art of yacht building and equipment that entitles the learner to usurp the position of teacher.

The following details of the dimensions of the rig, sail area of the contending yachts, will be read with interest by the initiated. For the information about the *Galatea* I am indebted to the courtesy of Mr. J. Beavor Webb, and the figures referring to the *Mayflower* were kindly furnished me by Mr. Burgess at the request of her owner, General Paine:

GALATEA — CUTTER.

Length over all	102.60 ft.
" of L. W. L.	87.00 "
" " lead keel	39.00 "
Beam extreme	15.00 "
Number of beams to length	5.80
Draft	13.¼ ft.
Mast deck to hounds	53.00 "
Topmast fid to pin	45.50 "
Boom extreme	73.00 "
Gaff pin to bolt	44.50 "
Bowsprit outboard	35.50 "
" close reefed	21.50 "
Spinnaker boom	66.00 "
Weight of lead keel	81.50 tons

AREA OF SAILS.

Mainsail	3,321 sq. ft.
Club topsail	1,365 "
Staysail	825 "
Jib	975 "
Jib topsail	1,265 "
Spinnaker	30,52 "
Bowsprit spinnaker or balloon jib topsail	2,530 "

MAYFLOWER — CENTERBOARD SLOOP.

Length over all	100 ft.
Length on L. W. L.	85½ "
Length of keel	80 "
Beam extreme	23½ "
Number of beams to length	3.6
Draft without centerboard	9¾ ft.
Draft with centerboard down	20 "
Mast deck to hounds	62 "
Topmast to topmast rigging	42 "
Total length of sticks from deck to truck	109 "
Bowsprit (which does not reef)	38 "
Main boom	80 "
Gaff	50 "
Spinnaker boom	67 "

AREA OF SAILS.

Mainsail	4,000 sq. ft.
Working topsail	800 "
Staysail	800 "
Jib	1,200 "
Spinnaker	4,000 "

So that when beating to windward the *Galatea* carried 7,751 square feet of canvas, while the *Mayflower* had approximately about 8,500.

THE SECOND RACE, SEPTEMBER 9.

As I threaded my way to the bows of the members' boat of the New York Yacht Club, on which Mr. Hurst, the treasurer, had kindly secured me a passage, I felt that I was about to witness the same performance outside the Hook as had saddened my spirits on the first day.

The weather was most unfavorable; drizzling rain commenced before we left Pier No. 1 and continued without intermission to speak of throughout the entire day. Added to these discomforts, a dense fog settled down early in the afternoon and put an end to the race and to any enjoyment of the trip, and sent us home groping our way, and landed us late, hungry and thoroughly miserable. In discussing this abortive attempt to finish this series of races I shall confine myself strictly to the details and technicalities of the contest, leaving the reader to supplement the accompaniments and accessories from my previous description, his vivid imagination or the details to be gathered from the voluminous expressions of opinion in the daily press accounts. The wind had risen considerably by the time we reached the Scotland lightship, and the weather gave angry tokens of letting loose a regular sou'wester. It was manifestly a clinking "cutter day," and right merrily did the *Galatea* lads move smartly about, taking a reef in the running bobstay, running in her bowsprit, hauling down the big jib, and substituting the second-sized one. Lieutenant Henn did not mean to be caught napping.

No change was made on the *Mayflower*. She carried her big jib and gained a great advantage thereby. Both craft thought it best to carry only working topsails.

At 11.20 the preparatory whistle was blown from the steam tug *Luckenback*, while the *Scandanavian* had been started ahead to mark out a twenty-mile course east by north, dead in the teeth of a fresh breeze of wind that put the racing craft scuppers to and sent the black waves seething and boiling in their wake.

Almost immediately after the starting signal the *Mayflower* bounded across the line, just skinning past the lightship. The *Galatea* was quite a good deal to leeward, and had to shake up a trifle into the wind to pass the judge's boat. Time of crossing was 11:30:30, and 11:30:32. Both craft were being sailed a shade fine, but the Boston sloop evidently held her way better, while the cutter made more leeway than she ought.

The *Galatea* did not relish her position, and at 11.50 made her first tack, quickly followed by the sloop. It was at once apparent that the old game had commenced, and the Boston boat, like a giddy girl, was romping away from her more sedate English sister. The difference in set of the sails of the two boats was also very noticeable, for while the *Mayflower's* canvas was stretched flat as a board, the leech of the *Galatea* kept licking about the whole way to windward, and must have been as annoying to her owner as it was disheartening to the gazing cutter men.

At 12.20 Sandy Hook lightship was passed, and the sloop had a clear lead of half a mile. The *Mayflower* made another short board at 12.58, returning to her original tack at 1.11. The Englishman held straight on. The wind shows a tendency to lighten, and at 1.27 the *Galatea* sent down her working topsail, and replaced it smartly by her club-topsail.

When about half the windward course was done the *Mayflower* appeared about 2½ miles distant, dead to windward of the cutter. At 1.37 the sloop tacked, and while shaking "in stays" her crew very smartly sent aloft her club-topsail to windward of her working one. The *Galatea* tacked again at 1.39, and apparently got a better wind, and seemed to have closed up the gap somewhat. At 1.50 the wind had lightened enough to allow the sloop to send up her jib-topsail. The sea also became smoother, and the fog began to settle down so thick that it was with difficulty the *Galatea* could be discerned a full three miles to leeward, which the sloop gradually widened to four or five before she rounded the mark buoy at 4:24:45 by my time. I saw nothing more of the *Galatea* that day, but read that she bore up for home when the *Mayflower* rounded. Fog, light wind, and closing darkness put an end to the race, which counted for nothing, as it was not sailed in the seven-hour limit, but it proved to the most skeptical the marked superiority of the sloop at the very game that was fondly believed to be *par excellence* a cutter's, for the *Mayflower* gained almost all her vantage while the sea and wind held. She outwinded and outspeeded the English cutter, and did not make nearly the leeway the *Galatea* did.

THE THIRD AND CONCLUSIVE RACE.
SEPTEMBER 11.

A glorious yachting day, a bright sun and a fresh steady breeze ushered in the final discomfiture of the cutter and her partisans. Space does not permit me to go into the details of the struggle; nor is it needed. The programme of Tuesday and Thursday was enacted without a hitch. The *Mayflower* left the *Galatea* in the run to leeward, increased the lead in the thrash to windward back home, and finally won the deciding event by 29m. 9s. For the subjoined history of the *America's* Cup I am indebted to my friend, Captain Roland F. Coffin, famous as a sailor, and still more so as the historian of sailors' deeds :

The cup which has once more been successfully defended by an American yacht, was first won by the schooner *America* in 1851, in a race of the Royal Yacht Squadron around the Isle of Wight, she sailing as one of a large fleet of schooners and cutters. The popular impression is that she sailed against the whole fleet ; but this is incorrect. She simply sailed as one of them, each one striving to win. When won it became the property of the owners of the *America*, and was brought by them to this country and retained in their possession for several years. They then concluded to make of it an international challenge cup, and by a deed of gift placed it in the custody of the New York Yacht Club as trustee. By this deed of gift any foreign yacht may compete for it upon giving six months' notice, and is entitled to one race over the New York Yacht Club course. There is, however, a clause in the deed which permits the challenger and the club to make any conditions they choose for the contest, and as a matter of fact, it has never been sailed for under the terms expressed in the deed of gift ; the two parties having always been able to agree upon other conditions.

When the schooner yacht *Cambria* came for it in 1870, she being the first challenger, the six months' notice was waived, and she sailed against the whole fleet, against the protest of her owner, Mr. James Ashbury, he contending that only a single vessel should be matched against her. The *Cambria* was beaten, and Mr. Ashbury had the schooner *Livonia* built expressly to challenge for this cup. The matter of his protest having been referred to Mr. George L. Schuyler, the only one of the owners of the *America* who was living, he decided that Mr. Ashbury's interpretation of the deed of gift was correct, and that such was the intention of the donors of the cup. When the *Livonia* came, in 1871, the club selected four schooners, the keel boats *Sappho* and *Dauntless*, and the centerboards *Palmer* and *Columbia*, to defend the cup, claiming the right to name either of those four on the morning of each race. The series of races was seven, the best four to win. There were five races sailed, the *Columbia* winning two, the *Sappho* two, and the *Livonia* one.

The next challenger was the Canadian schooner *Countess of Dufferin*, in 1876, and Major Gifford, who represented her owners, objected to the naming of more than one yacht by the New York club, and asked that she be named in advance. The New York club has from the first behaved in the most liberal and sportsmanlike manner in relation to this cup, and on this occasion it assented to Major Gifford's request and named the schooner *Madeleine*. The races agreed upon were three, best two to win. Only two were sailed, Capt. " Joe " Elsworth sailing the Canadian yacht in the second race. The *Madeleine* won both races with ease.

In 1881 a challenge was received from the Bay of Quinte Yacht Club, naming the sloop *Atalanta*, and the conditions agreed upon were the same as in the race with the other Canadian yacht, the club naming the sloop *Mischief*, which won the first two races.

The next challenger was the cutter *Genesta* last year, practically the same conditions being agreed upon as in the two previous races. The only difference was that as a concession to the challenger, two out of the three races were agreed upon to be sailed outside the Hook. The *Puritan* won the two first races, as the *Mayflower* has won them this year. From first to last, the only victory of either of the challengers has been that of the *Livonia* over the *Columbia*, which was gained by the American yacht carrying away part of her steering gear.

AMERICAN STEAM YACHTING.

AMERICAN STEAM YACHTING.[1]

BY EDWARD S. JAFFRAY.

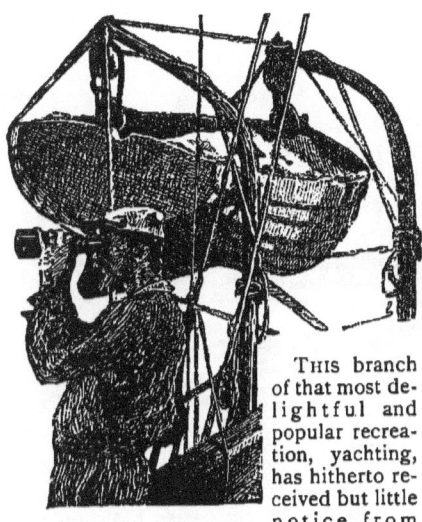

THIS branch of that most delightful and popular recreation, yachting, has hitherto received but little notice from writers on the subject. This is partly owing, no doubt, to the comparatively recent commencement of the use of steam in yachts.

As, however, this power is rapidly growing in favor, while sails remain almost stationary, it is desirable to place steam yachts in their proper place before the public, and give them, at least, a share of the attention and commendation which have hitherto been devoted almost exclusively to sails.

Your regular old yachtsman has a profound contempt for steam yachts. He considers that all the romance and pleasure of yachting consist in the uncertainties, dangers, and difficulties attending sailing. He glories in the storms which compel the shortening of sail, the lying-to, the scudding before the wind under a staysail, and all the other vicissitudes which attend excess of wind; while, on the other hand, he takes dead calms, with sails idly flapping against the masts, and the reflection of his vessel in the mirror-like water, with philosophy and contentment, passing the long hours of inaction in spinning yarns and (possibly) drinking cocktails. This class of yachtsmen is slowly passing away, and is being succeeded by men of more modern views. Gradually we see some of these gentlemen disposing of their sloops and schooners and ordering steamers to replace them. I can mention a few of these as an illustration — Commodore Bennett, Mr. William Astor, and Mr. Stillman.

The great truth is gradually dawning on the minds of yachtsmen that steam is the perfect motive power. Steam yachtsmen can go where they please and when they please, and, what is more important, *they know when they will get back.*

In this happy country, we are nearly all men of business, and we have neither time nor inclination to be becalmed on the glassy ocean for hours and days, or to creep along at three knots indefinitely. What the American yachtsmen require imperatively is the power of getting about with speed and certainty. With a steam engine on board, a man is able to command time and space, and is independent of storms and calms. "There is a tide in the affairs of men which, taken at the flood, leads on to fortune." This tide is *steam,* and though it may not always lead on to fortune, it invariably leads to the place whither the owner wants to go.

I have had a steam yacht for ten years, and in that time have traveled in it 55,000 miles, and as I have been constantly in company with sailing yachts, observing their picturesqueness and their helplessness, I think I am qualified to pronounce an opinion on the comparative merits of the two classes of yachts.

I may give a few instances of my experience. Some years since, as I was nearing Irvington,[2] on my return from the city, I met Mr. Stillman's yacht *Wanderer,* which had just got under way for a cruise. We exchanged salutes, and I went home. Next morning, on my trip down the river, I again encountered the *Wanderer,* which, after sailing (or floating) all night, had not yet reached Yonkers.[3]

On another occasion, I met the *Active,* Captain Hurst, at Thirty-fourth street, bound up the river. I proceeded to Twenty-third street, disembarked, went to my office, and, in the afternoon, at 3:30, started on my

---
[1] We desire to express our indebtedness to Mr. Charles Miller, of Nassau street, for permitting our artist to use many of his excellent photographs in the preparation of this article.
[2] Twenty-five miles from New York.
[3] Eighteen miles from New York.

"CORSAIR," OWNED BY J. PIERPONT MORGAN, OF HIGHLAND FALLS ON HUDSON.

usual homeward trip, and arrived at Irvington just as the *Active* was dropping her anchor. Her time from Staten Island, 33 miles, was about 12 hours.

In the race around Long Island, in 1884, there were fourteen of our best sailing yachts entered. I went to meet them at Execution Light, and arrived there just as the two winning boats, the *Grayling*, and the *Fanny*, hove in sight. Their time around the island was fairly good, but they were stopped a couple of miles from the stake boat by a dead calm, and I lay there two hours, while they were making their last two miles.

In the highly-interesting cruise of the New York Yacht Club, last summer, involving the trial races between the *Puritan* and *Priscilla*, the advantages of steam shone out conspicuously. The steamers were able to take any position they preferred, and thus, on leaving New London, they allowed all the sailing yachts to start, and then followed them, under easy steam; running along the whole extended line of schooners and sloops, viewing them from the most advantageous points, and running past them all in turn, until they reached the leading boats, which were the two champions, the *Puritan* and *Priscilla*, and they were then able to keep along with these at a short distance to leeward all the way to Newport.

Steam yachts may be divided into four classes. First, the launch, from forty to sixty feet long — an open vessel without deck; delightful vessels for river and harbor navigation.[1] The Herreshoff Company, of Bristol, R.I., have been the most successful in building this class of vessels, their launches showing a speed of ten to fourteen miles an hour. Some of these, like the *Camilla*, owned by Mr. Brandreth, of Sing Sing, and the *Lucille*, belonging to Mr. Herreshoff, are beautiful vessels, perfect in all their proportions, and of speed which enables them to perform runs of fifty to 100 miles in an afternoon. They have comfortable cabins, with glass windows, in which their occupants can enjoy the scenery, while completely protected from the weather; and for use on the Hudson River and similar waters they are all that could be desired.

In the second class I put regular decked vessels of 75 to 100 feet long, which have trunk cabins. They have not depth enough to have cabins with a flush deck above them, and therefore the deck, which is "par excellence" the best part of the vessel, is sacrificed to the cabin. As yachting is carried on only in the summer (as a rule), when it is pleasant to be in the open air, yachtsmen and their guests are always on deck, viewing the scenery and the passing vessels, except when the announcement by the steward that a meal is ready causes them to hurry down to the saloon with generally, I presume, excellent appetites. As soon, however, as the eating

[1] I omit all launches below forty feet in length, as the New York Yacht Club does not recognize any vessel of less than forty feet long as a yacht, and does not admit them into the club.

"CAMILLA," OWNED BY COL. FRANK BRANDRETH, OF SING SING, N.Y.
(Drawn by Cozzens, New York and American Yacht Clubs.)

is accomplished, they return to the deck to smoke their cigars and see what is going on. Now, to sacrifice the deck merely to have a more roomy cabin, is, I consider, a fatal mistake, and I consequently disapprove in toto of this class of vessels. Beside the loss of the deck there is another serious objection to them. They are not safe in a sea way. A sea taken on board might easily crush in the sides of the trunk cabin and swamp the vessel, and, consequently, these yachts are not fit to go into the open sea except when the barometer stands above thirty and the ocean is in a quiet mood.

The third class consists of vessels somewhat larger than the preceding, and especially having greater depth, with a flush deck from stem to stern. These yachts are very desirable, and can go anywhere. Among the best of these are the *Pastime*, the *Sentinel*, and the *Tilley*. This class of yachts should satisfy all persons who propose to navigate the Hudson River, the Sound, and as far along the coast as Mount Desert Island, where they can make a port at the end of each day's run, and do not require to pass the night on the open sea.

The fourth class consists of larger vessels, which are regular sea-going craft, fit to circumnavigate the globe. Of such are the *Nourmahal, Namouna, Atalanta*, and I may add, though they are of a somewhat smaller class, the *Electra*, the *Corsair*, and the

HASSAN STEAM LAUNCH OF JAMES GORDON BENNETT.
(From a drawing by Hennelle, in *Yacht*.)

Stranger. These all have flush decks of ample dimensions, and large saloons and state-rooms, and in fact combine all the qualities necessary to make them the perfection of comfort and pleasure.

There is probably no better yachting ground than the waters around New York and the coast of New England, as far as the Bay of Fundy. For 200 miles, with the exception of the run from Watch Hill to Cuttyhunk, the waters are protected by outlying islands. The voyage, then, from Oak Bluffs to Portland is in the open sea for three-fourths of the distance, but from Portland to Bar Harbor the navigation again is in inland waters, so that in the cruise of 550 miles the course exposed to the open sea is not more than 200.

While sailing yachts have a troublesome and difficult navigation through Nantucket Shoals to reach Cape Cod, steamers can lie the direct course from light ship to light ship, feeling their way along, guided by the bell or fog whistle of the various light vessels, and can navigate with comparative safety and certainty through the labyrinth of sand banks, while the sailing yachts, baffled by light winds, and embarrassed by fogs, have to anchor or turn back till a favorable change in the weather.

A good steamer, with a speed of 15 to 17 miles an hour, can make this eastern cruise about as follows. First day run to that delightful harbor, New London, 110 miles ; next day to Newport, 46 miles ; and after staying a couple of days for the festivities and hospitalities sure to be found there, run to Oak Bluffs, about 50 miles. A day there will suffice to see the thousand ornamental cottages, after which, starting at daylight, run to Portland, 190 miles, going through the Shoals, and skirting the long, sandy shore of Cape Cod, passing in turn the various life-saving stations and light-houses, and, after reaching the end of the promontory making a course almost due north, to the fine harbor of Portland.

A delightful excursion may be made, while here, to the head of Casco Bay, some thirty nautical miles, running up one avenue of beautiful verdure-clad islands, returning down another equally interesting. The next run should be to that charming spot Bar Harbor, about 120 miles direct; but the distance may be increased to 160 by going in and out among the crowd of picturesque islands, and following the line of the undulating shore. The yacht would thus pass close to Rockland, ~~Rockford~~ *Rockport*, and Camden,

"UTOWANA," OWNED BY W. E. CONNOR, REAR COMMODORE BOSTON YACHT CLUB.

in the beautiful bay of the latter name which has as a background the fine range of the Camden Hills.

From Bar Harbor the cruise may be continued to Campo Bello, to St. John, to Halifax, and round into the Gulf of St. Lawrence, as may be most agreeable.

With a steam yacht of the larger class one may do *anything*. There are no limits to the enjoyments of such a mode of traveling, and when it is desired to return, one may telegraph the exact day, and almost the hour, when he will drop his anchor again in the Hudson.

As a compromise between steam and sails, I would suggest the sailing vessel with an auxiliary screw, like Sir Thomas Brassey's yacht, the *Sunbeam*. In this he has circumnavigated the globe, and cruised in the Mediterranean many times with great success and comfort, and there is no more agreeable reading to be found than Lady Brassey's graphic accounts of these voyages. This class of vessel combines the delightful romance and uncertainties of the sailing yacht with the power to get through calms and against head winds, when necessary, by means of the steam engine. It can thus go on long voyages without the inconvenient burden of a large cargo of coal, as in vessels propelled wholly by steam, and all the interesting experience of navigation by sails can be enjoyed for weeks and months together, so that one might almost forget the engine and boiler down below, and feel as if the winds were the only propelling power. For long voyages this combination of sails and steam has undoubtedly great advantages, being superior to either style alone. But there are very few yachtsmen who have the leisure or the desire to go off on a six months' cruise, and for all river and harbor and coasting expeditions the steamer is the true style of vessel. The *Sunbeam*, which, I presume, is one of the most successful of her class, and a perfectly satisfactory vessel to her owner, would cut but a poor figure in a run up the Hudson or the Sound in company with our better class of yachts. The best speed of the *Sunbeam* under steam alone is, I believe, 8 knots, while our steamers run from 10 to 17 knots, so that, starting in company, as I have supposed, she would be out of sight

"ATALANTA," OWNED BY JAY GOULD, AMERICAN YACHT CLUB.

astern in a couple of hours run. There is nothing so galling to a man of fine feelings, when yachting, as to have another yacht come up and go past him. Under these circumstances a man is tempted to sit on the safety valve and turn on the steam jet, burn rosin and kerosene, and to do anything desperate to avoid such a humiliation; and this spirit of competition and emulation is one of the greatest helps to the development of excellence in building these vessels. Every man who gives an order for a steam yacht directs the builder to make it a little faster than any previous vessel, and thus the ingenuity of the enterprising builders is taxed to the uttermost, and excellence is the natural result. There is, however, a limit to the speed of such vessels; every additional mile added to the speed is only obtained by an enormously increased power and expenditure of fuel, and when a certain speed is reached (dependent, of course, somewhat on the model of the vessel), the resistance and slip balance the power employed, and no further increase can be obtained. Our American yachts make better time, as a rule, than those of England, the latter seldom attaining a greater speed than 10 knots, while our larger class make from 12½ to 17. The *Atalanta* can steam 17 knots, the *Corsair* and *Stranger* 15, and a number of others 14, thus showing either that our models are the center of the boat, working in an airtight iron box, into which air was forced, for the purpose of keeping the water down.

The invention proved a failure, and then it was that Mr. Aspinwall altered this boat by putting on her side-wheels with feathering buckets, and an oscillating engine, and thus produced, so far as there is any record, the first steam yacht in New York harbor, and probably the first in America. She was named the *Fire-Fly*, and he used to come up in her quite frequently to his business in New York from his country-seat

"PASTIME," OWNED BY R. C. WALKER, OF DETROIT, MICH.

better, or that we use engines of greater power.

I will now give a sketch of the rise and progress of steam yachting in this country, and in doing so, I tender my acknowledgments to the Rev. John A. Aspinwall and Mr. Jacob Lorillard for much valuable information which they have kindly furnished.

About thirty-three years ago, Mr. William H. Aspinwall, of New York, the President of the Pacific Mail Steamship Company, built a steam-boat 50 or 60 feet long, in order to try an experiment with a wheel, which a Frenchman had invented, and which it was thought would be a success. It consisted of a single paddle-wheel, in on Staten Island, and take pleasure-trips down the bay and sound. Her captain was named, Dayton, and her engineer, John Armstrong. Her speed was from nine to ten miles an hour.

This boat was afterwards bought by the Government, and went south at the beginning of the civil war.

Mr. Aspinwall then built his second boat, the *Day-Dream*. She was a composite vessel, 105 feet on the water-line; 17 or 18 feet beam. She had a pair of upright engines in her, and one horizontal boiler, and an inside surface condenser.

The model was made by Dr. Smith, of Green Point, and under his supervision the

"SENTINEL," OWNED BY J. L. ASPINWALL, ATLANTIC YACHT CLUB.

boat was built, at the Continental Iron Works. Her machinery was constructed by the Delamater Iron Works, of New York. The speed of this vessel was from twelve to fourteen miles.

About the time that Mr. Aspinwall built his first steam yacht, the *Fire-Fly*, his son John, then a school-boy of about thirteen years old, having a natural taste for mechanics, used to go Saturdays to the Morgan Iron Works and the Allaire Works, and became very much interested in the construction of boats, boilers, and engines. He therefore determined to try and build himself a steam yacht, to be used on a pond, on his father's place. He succeeded in building a flat-bottom boat, with a sharp bow, twelve feet long, and three feet six inches wide, and placing in her an engine driven by six alcohol lamps, and attaching it by cog-wheels to the shaft. He paddled about in the pond at the rapid speed of half a mile an hour, to the great consternation of the two stately swans, the discomfiture of the large frogs, sitting on the floating bits of wood, and the rapid diminution of the alcohol, gotten, from the demijohn in his father's pantry. As a matter of fact, in all his after-experience in steam yachting he has never been able to reduce the item of fuel down to so low a figure as in his first experiment.

Two or three boats followed this first one, all of them built by himself and playmates. In 1865, the year after he went to reside in Bay Ridge, L.I., and take charge of the Episcopal Church and parish there, he, as a pastime and recreation, and for the purpose of getting a stronger hold on some of the young men in the parish, built, together with them, a flat-bottomed, side-wheeled boat about 20 feet long, drawing only 5 inches. This boat had been in the water a short time, when a Southerner, from Savannah, who offered just twice what she cost, bought her and took her to the Savannah River. Then Mr. Aspinwall had built, by a Mr. Whitman, in South Brooklyn, a side-wheeled boat called the *Julia*, about 35 feet long. The model was furnished by a friend. The boiler and engines were built by the Continental Works of Greenpoint.

This yacht had such a round bottom that on her first trial she was nearly upset, while crossing the East River, by the waves from a passing Sound boat. She never left the dock but that once. Her engine and boiler were put into another boat, about 45 feet in length, designed chiefly by her owner. She gave great satisfaction, and was called *Julia No. 2.* Then followed the *Comet*, a boat 25 feet long, the first one in which he put a propeller engine. Then came the *Surprise*, built by James Lennox, of South Brooklyn.

This yacht was 65 feet long, and had a pair of upright engines, an upright tubular boiler, and an outside pipe condenser, and was the first yacht in New York harbor that had a pilot-house. She was sold to Captain Billings, of New London, who still is her owner. Next came the *Runaway*, 70 feet long; after using this yacht two seasons, the owner cut her in two, just forward of the boiler, and lengthened her fifteen feet, and changed the boat from a trunk cabin to a flush deck. This work was done by a Mr. Voris, of Upper Nyack. This was the first flush deck steam yacht, owned in New York. She had a pair of upright boilers and engines, and made 10 to 12 miles per hour, with no change in her machinery; the addition made to her length having had no perceptible effect on the speed. She was purchased by General Newton for the Government, and is still doing service at Astoria. After this he built a boat called the *Arrow*, 70 feet long, 10 feet beam; her engines and boiler were like those in the *Surprise*, but generally improved in design and construction. This yacht attained the speed of 13 miles, and was at that time about the fastest boat of her size in the harbor. Mr. Aspinwall, of Barrytown, purchased her.

The *Pastime* was then built, having compound condensing engines, known as the tandem pattern, high-pressure cylinders, 10 inches by 12, low pressure 12 inches square, upright tubular boiler, with outside pipe condenser. This boat was sold to Mr. E. T. Gerry, of New York, who resold her, after he built the *Electra*, to parties who took her far from New York. In 1880, he built a small boat 34 feet long, 9 feet beam, with a pair of 4 by 4 engines and an upright boiler. This boat was called the *Dart*, and was sent south, where her owner now uses her for navigating the bays and rivers about St. Augustine.

Boat No. 13 was named the *Sentinel*, being the one which he at present owns, 106 feet on water-line, 18 feet beam, flush deck, pair of compound condensing engines, two high-pressure cylinders, 12 by 12, two low-pressure, 20 by 12, upright tubular boiler, containing 520 2-inch tubes, and capable of supplying engines at full speed (with blower running) with steam at 100 pounds pressure; speed of this yacht, 12 to 15 miles per hour. All the engines and boilers of the last seven yachts were built by Mr. Lysander Wright, Jr., of Newark, N.J., and have proved themselves to be most excellent specimens of workmanship and durability.

When Mr. J. A. Aspinwall, as a boy, first began to build boats, he made up his mind to understand, so far as possible, every department of a steam yacht, so that he has been, by turns, deck hand, cook, fireman, engineer, and captain. He ran the engines of all his own yachts until they were so large that he required more crew than he and his friends could supply; but up to the present time he has always been his own captain and pilot, having been licensed as such for the past ten years.

The next contributor to the steam-yacht fleet was Mr. Jacob Lorillard, who commenced building in the year 1868. This gentleman has done more to create and foster the fashion for steam yachting than any other person. His plan was to build a new yacht every year, and after using it through the season, and getting it into first-rate running order, to sell it, and thus make room for a new vessel the succeeding year. His yachts were thus transferred to other cities, and in this way they contributed to spread the fashion for this pastime. It has consequently now become a great national amusement, and is constantly growing in all our sea-board cities.

I insert here a letter from Mr. Lorillard,

"ELECTRA," OWNED BY ELBRIDGE T. GERRY, COMMODORE NEW YORK YACHT CLUB.

NAME.	Year Built.	Rig.	Length of Water-line.	Length on Deck.	Beam.	Depth.	Draught of Water.	Diameter of Screw with Best Result.	Pitch of Screw with Best Result.	Revolutions per Minute.	Diameter of Cylinder.	Stroke.	Pressure.	Type of Boiler.
				Ft.	Ft.	Ft.	Ft.	Ft.	Ft.		In.	In		
Firefly	1868	Schooner	62	67	13	5½	5	4½	9	140	16	12	40	Horizontal water tube.
Mischief	1869	"	71	76	15	5½	4½	4½	9	140	16	12	60	" "
Emily	1869	Brigantine	71	76	15	5½	4½	4½	9	140	16	12	60	" "
Fearless	1872	Schooner	80	85	15	5½	4½	4½	9	142	9-16	12	80	Upright tubular.
Lurline	1872	"	80	85	16	5½	5½	5	9½	142	13-22	16	90	Return horizontal tubular
Skylark	1874	"	78	84	16	5½	5	4½	9	145	12-20	12	100	Upright water tube.
Lookout	1875	"	100	105	16	5½	5	5	9½	160	12-20	16	100	Horizontal water tube.
Truant	1876	"	77	86	16	5	4½	4½	9	142	9-16	12	100	Field tube (hanging).
Promise	1877	"	91	98	16	5½	5	5	8½	230	12-20	16	130	" " "
Rival	1878	"	91	98	16	5½	5	4½	9	200	12-20	16	90	Scotch 7 feet diameter.
Theresa	1880	"	90	96	16	5½	5	4½	9	175	12-20	12	70	" "
Minnehaha	1875	"	57	60	8½	4	4	3	5	300	pair 8	8	100	Water tube.
Tillie	1882	"	105	119	17	5½	5½	5	9½	140	14-24	16	90	Scotch 9 feet diameter.
Winifred	1882	"	91	98	16	5½	5	4½	9	145	12-21	16	80	Scotch 6½ feet diameter.
Vision	1883	"	90	98	16	5½	5	4½	9	150	12-21	16	90	Field water tube.
Venture	1883	"	62	70	12½	4	3½	3	5	180	10	10	120	Upright tube.

The Theresa gave the best results of any, owing to form of model. Slip 3 per cent.
The Promise had excessive power to displacement and speed, as did Minnehaha (power models).
The Truant gave greatest economy for displacement and speed. Model good for limited speed. Slip 3½ per cent., only being minimum, except Theresa.
Rounding frames proved least resistance.
Lengths of 5½ to 1 beam gave least results.
Field boilers proved fastest evaporators and lightest, but were unsatisfactory (ends of tubes burning) evaporated double per square foot surface of any others.

with a full list and description of the yachts he has built:

JANUARY 19th, 1886.

E. S. JAFFRAY, Esq:

DEAR SIR,—I herewith enclose you chart of the dimensions, power, displacements, etc., of some of the yachts I have built. On nearly all of them I have had a series of different screws of various sizes and pitches, and the ones specified were those that gave the best result after many trials. I weighed all the vessels in the screw dock to check calculated displacement, and the displacement given was within a ton or so of the actual weight, as the coal in them was estimated, and might vary a ton or two at most.

A careful study of the results, when the model is before you, will convince one of the desirability of curved lines in every direction, and particularly so in the sectional lines. The midship body was very near the center of the boat in all that gave speed over 13 miles, and I am convinced that this must be moved forward as the rate of speed increases on a vessel of a given length. I am now building one 93x16, that has midship frame 3½ feet forward of the center of keel. I have filled up both hollow ends to make as regular an arc from end to end as possible, and with rounding frames expect to get a minimum of resistance, and good speed and economy of power.

Very truly yours, etc.,

[Signed.] JACOB LORILLARD.

The principal builders of steam yachts in America are Cramp & Sons, of Philadelphia; The Harlan & Hollingsworth Company, of Wilmington, Del.; Mr. John Roach; The Herreshoff Company of Bristol, R.I.; and Ward, Stanton & Co., of Newburgh (now defunct). Messrs. Wm. Cramp & Sons describe the four yachts that they have built, in answer to a letter from me on the subject:—

"In compliance with your wish expressed in your favor of the 12th inst., the following-named steam yachts: *Atalanta*, *Corsair*, *Stranger*, and "246," are the only ones we have built, and all have been very successful, each one not only coming up to the speed required, but going in every case beyond.

"*The Stranger* was built for Mr. George Osgood, of New York, in the year 1881; she was (up to the time that the *Atalanta* was built) the fastest yacht in America. Mr. Osgood, on one occasion, took his breakfast at Newport, at 7 A.M., and at 4 P.M. the same day was taking his dinner in New York harbor, having made the run of 135 knots in nine hours, being 15 knots per hour. This is the best time made between Newport and New York, either by a side-wheeler or propeller. The *Corsair*

Name.	Average Displacement in Tons.	H. P.	Average Speed in Still Water.	Memoranda.
Firefly	20	60	14½	Had trunk cabin 30 inches high above deck.
Mischief	30	75	13½	Trunk cabin, one state-room, scag 1 foot below keel.
Emily	31	75	13½	" " " " " lengthened (1870) 13 feet, scag 1 foot.
Fearless	29	60	12	Compound engines, altered from single, show gain by test of 41 per cent. of fuel.
Lurline	45	130	14½	Trunk cabin, scag 15 inches below keel; lengthened 13 feet in 1873.
Skylark	40	85	14	" " " " " " " " "
Lookout	45	140	14½	" " " " " " " " "
Truant	21	50	13	" " " " " " " " "
Promise	40	270	16	" " " " " 18 " " "
Rival	45	200	14½	Boiler forced by blower to burn 40 lbs. per square foot grate evaporative, 1 cubic foot water per 4½ square feet heating surface.
Theresa	40	90	16½	" " " " " " " " "
Minnehaha	10	60	15	" " " " " " " " "
Tillie	50	140	14	Lengthened 30 feet in 1884 (power too small for new hull), with insufficient steaming capacity.
Winifred	45	90	13½	Trunk cabin.
Vision	45	100	14½	" " (boiler replaced with upright tubular).
Venture	14	50	12	" "

Compound engines saved 41 per cent. of coal over single engines at same point of expansion.
Scotch boilers were very heavy for quantity of steam produced.
For small yachts would recommend upright boilers as best for natural draught boats.
None of the above yachts were built entirely for speed, but to combine good speed with comfortable accommodations and an economy of fuel for cruising purposes, and to be graceful vessels.
A light and compact boiler that will produce a large evaporation is greatly wanted for yacht purposes.

was equally fast, both vessels having been of the same model, and supplied with the same engines and boilers. They are 185 feet long, 23 feet beam, and 13 feet hold, and 9 feet 3½ inches draft of water, compound engines, high-pressure cylinder 24 inches diameter, low-pressure cylinder 44 inches diameter and 24 inches stroke, developing 760 horse power. The only difference between the two yachts was a different finish about the cutwater.

"The *Atalanta* was built, as you know, for Mr. Jay Gould, in the year of 1883, and has also come fully up to the expectations as to speed, comfort, and seaworthiness entertained by her owner and builders. She is a perfect model of beauty and comfort, and is a knot and a half or two knots faster than any other yacht in the world. She has made the speed of 20 miles in one hour and a quarter, with fire under but one of her boilers. When she has her full compliment of steam from both boilers, there is not a vessel of her inches that can keep alongside of her. She is magnificently fitted up with hardwood saloons and state-rooms. She is 248 feet 3 inches long, 26 feet 5 inches beam, and 16 feet depth of hold on a draught of 12 feet of water. Her engines are compound 30 inches diameter low-pressure cylinder by 30 inches stroke, 110 pounds steam, and indicates 1,750 horse power. She has attained a speed of 17 knots per hour.

"The yacht '246' is 166 feet long, 22 feet beam, 13 feet hold, with 8½ feet draft of water, engines triple expansion, 17 inches diameter high pressure, 24 inches diameter intermediate, and 40 inches diameter low-pressure cylinders, with a stroke of 22 inches. In the yacht race in July last, from Larchmont to New London (a ninety-five mile run), we came in behind the *Atalanta* eleven minutes. Being a new vessel, and just from the yard in an unfinished state, this shows an extraordinary performance for a yacht of this size."

The Harlan & Hollingsworth Company have kindly placed at my disposal descriptions of the six steam yachts they have turned out; two of them ranking among the finest yachts in the world.

"METEOR," 1876.

Dimensions: Length between perpendiculars, 75 feet; length over all, 79 feet 3 inches; breadth of beam, 10 feet; depth amidships, top of keel to top of beam, 5 feet 1 inch; mean draft, 3 feet; displacement (finished), 20 tons.

Machinery: Two inverted, high-pressure engines, 10x12 inches.

Boiler: Of steel, for 200 pounds pressure.

"VIKING," OWNED BY HON. SAMUEL J. TILDEN, OF GREYSTONE ON HUDSON.

Propeller: 56 inches diameter; 300 revolutions per minute.

Joinery: Saloon on main deck aft; pilot house forward.

Accommodations for officers and crew; capstan on main deck forward.

Speed: 21 miles per hour.

"VICTOR," 1878.

Dimensions: Length over all, 55 feet; beam, molded, 10 feet; beam, over guards, 13 feet; depth, amidships, 4 feet 6 inches.

Machinery: One direct-acting, vertical, surface-condensing engine; cylinder 6 inches diameter by 8-inch stroke, with attachments complete; donkey feed pump, injector, etc.

Boiler: Vertical, tubular boiler, 36x63 inches for a working pressure of 100 pounds per square inch; air jet into stack.

Propeller: 36 inches diameter.

Joinery: Cabin forward, with berths, lockers, etc., finished in hard wood; pantry, kitchen, etc., aft.

Two masts, schooner rigged.

"DIONE," 1879.

Dimensions: Length between perpendiculars, 43 feet; length over all, 47 feet 4 inches; beam, molded, 7 feet 9 inches; depth from base line to top of gunwale plate at dead flat, 3 feet 10 inches; depth from base line to top of gunwale plate at dead flat, at stem, 5 feet 4½ inches; depth from base line to top of gunwale plate at dead flat, at end of counter, 4 feet 7½ inches.

Machinery: Inverted, direct-acting, propeller engine 8x10 inches.

Joinery: Trunk cabin, with all accommodations for pleasure and comfort.

Boiler: Horizontal locomotive type of boiler for 120 pounds working pressure.

"FALCON," 1880 (IRON HULL).

Dimensions: Length between perpendiculars, 100 feet; length over all, on deck, 107 feet; breadth of beam, 15 feet 6 inches; depth from base line, 7 feet 6 inches.

Machinery: One vertical, direct-acting, condensing engine, cylinder 16 inches by 16 stroke. Propeller, 5 feet 7½ inches in diameter.

Boiler: One high-pressure boiler, with two furnaces, flues below and return through tubes; arranged for a working pressure of 100 pounds per square inch.

Joinery: Joiner's work of soft wood, except the dining room and social hall forward, and two state-rooms aft, which are finished in hard woods.

Forecastle forward, and two state-rooms under the social hall.

Pilot house on promenade deck forward, with room abaft of same. Promenade deck fitted with rail and stanchions with rope netting, also awning stanchions and awning frame.

Iron water tanks; hand fire pump; ice box; iron cranes for carrying boats; anchor crane forward; oil tanks, etc.

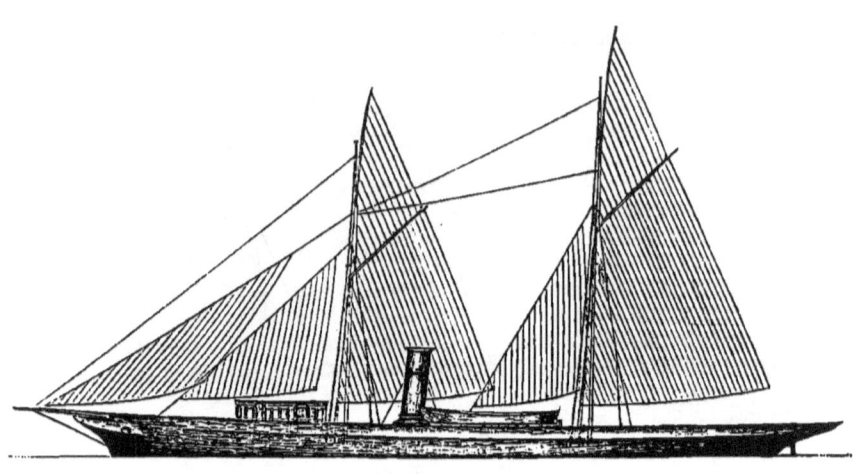

"WANDA," OWNED BY W. WOODWARD, JR., AND JAMES STILLMAN, EASTERN YACHT CLUB.

"NOOYA," J. H. ANDREWS, OWNER.

AMERICAN STEAM YACHTING.

STEEL YACHT "NOURMAHAL," 1884.

Dimensions: Length on deck foreside of rudder post to afterside of stem (or its rake line), 232 feet 5 inches; length on waterline, 221 feet; breadth extreme (or its rake line), 30 feet; depth of hold, top of floors to underside of deck, amidships, 18 feet 7½ inches; depth molded, top of keel to top of beams at sides and amidships, 20 feet; with five athwartship and two fore and aft bulkheads.

The *Nourmahal* is a queen among steam mann, of City Island, N.Y., who designed the vessel, superintended her construction; and an inspector of the English Lloyds, under the rules of which the yacht was built, performed his duty thoroughly. Frequent reference to the construction of the *Nourmahal* in these columns have contained in full the measurements and scientific data involved, which need not be repeated, but her internal arrangements and fittings are entitled to consideration. The *Nourmahal* looks the ocean yacht all over. Her model is exceedingly shapely,

"NOOYA"—DECK VIEW.

yachts. She is of steel throughout, and in construction and fittings neither time nor money have been considered. A pleasure vessel capable of any service, either under steam or canvass, was required, and it is believed there is not afloat, to-day, in any clime, a stronger, handsomer, or more perfect craft. Almost a year has elapsed since the Harlan & Hollingsworth Company, of Wilmington, Del., began the preliminary works incident to the building of this vessel, and had it not been for delays impossible to prevent, in the matter of obtaining required amounts of steel at required times, the *Nourmahal* would have been ready months ago. Mr. Gustav Hill- and the long, easy lines, with reduced area of amidship section, cannot fail to attract attention. The bow has a peculiarly rakish appearance, and her elliptical stern is very handsome; and while it is claimed there is greater strength in this construction, it is certainly less dangerous than the square stern when running before a heavy sea. The plating of the vessel above the waterline is smooth as a board, and the neat manner in which this work is done demands especial mention. The hull is painted a glossy black, with a female figure and a delicate gold tracery at the head as an ornamentation, and on the stern there is nothing but her name, and the port from

"PROMISE," OWNED BY A. D. CORDOVA — LARCHMONT YACHT CLUB.

which she hails, in massive gold letters. She has one large single smoke-pipe, also painted black, and she is bark rigged. Numerous large lights are on the sides for air and light, and eight coaling ports have been provided. Externally the *Nourmahal* is a yacht of grand proportions and rakish beauty, capable of all around the world explorations, and of strength sufficient to laugh at the fitful moods of the ocean. Internally there is a world of room, supplied with every known novelty of approved excellence, while the finish, fittings and decorations are of a very costly nature and magnificent in their exquisite simplicity.

The *Electra* was designed by Mr. Hillmann to combine strength, speed and convenience, and she was built under the superintendence of Mr. Gerry and the architect, under the rules of the English Lloyds. Her dimensions are as follows: deck length, 178½ feet; at water-line, 161 feet; beam, 23 feet; hold, 13¼ feet; draught, 9½ feet. She is built in the stanchest manner. Her motive power is a propeller 8 feet in diameter, with 13 feet pitch, capable of 160 revolutions a minute. It is turned by an inverted direct-acting compound engine, with high-pressure cylinder 22 inches in diameter, and a low pressure cylinder of 40 inches diameter and a stroke of piston 26 inches. Steam is generated in two cylindrical steel shell boilers, each 11 feet long, 10½ feet in diameter and supplied with furnaces 42 inches in diameter. The engines and boiler rooms occupy the whole width of the vessel and occupy a space 50 feet in length, with coal bunkers on either side and under the forward cabin capable of carrying 100 tons. The smokestack is double, and so arranged that a pipe from the kitchen connecting therewith carries off all the smoke and smell of cookery.

In the engine room are also located the engines for running the fifty-eight Edison electric lights of 16-candle power each, by which the boat is mainly lighted, as well as the Edison light of 100-candle power at the mast-head and the side electric signals; the ice machine, which makes 56 pounds a day; an independent condenser, not connected with the frame of the engine; independent air, circulating and feed pumps, as well as an independent steam fire and bilge pump, and a blower to blow into an air-tight fire room and to aid in the proper ventilation of the cabins.

She has six water-tight bulkheads, and all the connecting doors shut water-tight; in addition to her steam propelling power, she has a schooner rig, with top masts, and carries a forestaysail jib, foresail, two gaff topsails and a spanker. She also carries four boats, including a gig 24 feet long, a life boat 21 feet long, and two dingheys each 17 feet long. Her gross registry of tonnage is 303.98 tons.

The following is a list of the steam yachts built by the Herreshoff Manufacturing Company of Bristol, R.I:

Aïda, built 1882, for Mr. Mark Hopkins,

St. Clair, Mich. Length, 95 feet; breadth, 12 feet 6 inches; depth, 6 feet 3 inches; draught, 4 feet 6 inches; speed, 16 miles per hour.

Camilla, built 1881, for Dr. J. G. Holland. Length, 60 feet; breath, 9 feet; depth, 4 feet 7 inches; draught, 3 feet, 5 inches; speed, 15 miles per hour.

Dolphin, built 1879, for Robert Lenox Kennedy. Length, 42 feet; breadth, 8 feet 6 inches; depth, 4 feet; draught, 3 feet; speed, 10 miles per hour.

Edith, built 1880, for William Wood-

Magnolia, built 1883, for Fairman Rogers, Philadelphia. Length, 99 feet; breadth, 17 feet 6 inches; depth, 8 feet 6 inches; draught, 4 feet; speed, 11½ miles per hour. This vessel has twin screws, and is the only yacht of this kind in the United States.

Nereid, built 1882, for Jay C. Smith, Utica, N.Y. Length, 76 feet; breadth, 12 feet 6 inches; depth, 6 feet 3 inches; draught, 4 feet 6 inches; speed, 14 miles per hour.

Orienta, built 1882, for J. A. Bostwick,

"SUNBEAM"—SIR THOMAS BRASSEY, OWNER.

ward, Jr., New York. Length, 60 feet; breadth, 9 feet 2 inches; depth, 4 feet 7 inches; draught, 3 feet 5 inches; speed, 15 miles per hour.

Gleam, built 1880, for William H. Graham, Baltimore. Length, 120 feet; breadth, 16 feet; depth, 6 feet 5 inches; draught, 5 feet 8 inches; speed, 17 miles per hour.

Idle Hour, built 1879, for B. F. Carver. Length, 60 feet; breadth, 9 feet; depth, 4 feet 7 inches; draught, 3 feet 5 inches; speed, 15 miles per hour.

Juliet, built 1881, for Morris & Jones, Bartom-on-the-Sound. Length, 45 feet; breadth, 9 feet; depth, 4 feet 3 inches; draught, 3 feet; speed, 11 miles per hour.

New York. Length, 125 feet; breadth, 17 feet; depth, 8 feet 6 inches; draught, 6 feet 6 inches; speed, 17 miles per hour.

Ossabaw, built 1883, for Archibald Rogers, New York. Length, 69 feet; breadth, 9 feet; depth, 5 feet; draught, 3 feet 6 inches; speed, 16 miles per hour.

Permelia, built 1883, for Mark Hopkins, St. Clair, Mich. Length, 100 feet; breadth, 12 feet 6 inches; depth, 6 feet 6 inches; draught, 4 feet 6 inches; speed, 19½ miles per hour.

Siesta, built 1882, for H. H. Warner, Rochester. Length, 98 feet; breadth, 17 feet; depth, 8 feet 6 inches; draught, 5 feet 6 inches, speed, 13½ miles per hour.

Sinbad, built in 1879, for F. S. de Hauteville, New York. Length, 42 feet; breadth, 8 feet 8 inches; depth, 3 feet 9 inches; draught, 3 feet 2 inches; speed, 10 miles per hour.

Speedwell, built 1876, for Walter Langdon. Length, 45 feet; breadth, 6 feet 9 inches; depth, 3 feet 3 inches; draught, 2 feet 8 inches; speed, 12 miles per hour.

Sport, built 1880, for Joseph P. Earl, New York. Length, 45 feet; breadth, 8 feet 2 inches; depth, 3 feet 2 inches; draught, 1 foot 2 inches; speed, 10 miles per hour.

Marina, built 1884, for G. A. Bech, Poughkeepsie, New York. Length, 87 feet;

PAINTING THE BOAT.

breadth, 12 feet 6 inches; depth, 7 feet 3 inches; draught, 5 feet; speed, 14 miles per hour.

Leila, 100 feet by 15 feet, built 1887, for William H. Graham, of Baltimore. One engine, 9 inches and 16 inches by 18 inches; Herreshoff coil boiler 6½ feet diameter; speed, 16 miles an hour.

Kelpie, 47 feet by 7 feet, built 1878 for William H. Graham, of Baltimore. Engine, 3½ inches and 6 inches by 7 inches; boiler, 42 inches diameter, Herreshoff coil; speed, 12 miles.

Lucy, same size and description as *Dolphin*, built for F. S. Birch, New York.

Lucille, 69 feet by 9 feet, built 1884 for Charles Kellogg, of Athens, Penn. Engine, 6 inches and 10½ inches by 10 inches; boiler, Herreshoff patent safety 56 inch square; speed, 19 miles.

Polly, duplicate of the above *Lucille*, except speed, which was 17 miles; built 1885 for C. A. Whittier, of Boston, Mass.

Lucille, 90 feet by 11½ feet, engine 8 inches and 14 inches by 14 inches; boiler, Herreshoff safety, 67 inches square; built 1885 for Charles Kellogg, Athens, Penn.; speed, 17 miles.

Ladoga, 97 feet by 13 feet; engine, 8 inches and 14 inches by 14 inches; boiler, Herreshoff safety, 67 inches square; built for George Gordon King, of Newport, R. I., 1885.

Augusta, 55 feet by 6½ feet, side wheel; engine, 6 inches by 24 inches; boiler, Herreshoff coil, 42 inches diameter; built 1882, for Charles Kellogg of Athens, Penn.; speed, 14 miles.

Stiletto, 94 feet by 11 feet, built for "H. M. Co."; engine, 12 inches and 21 inches by 12 inches; boiler, Herreshoff patent safety, 7 feet square; speed 25 miles; built 1885.

In the year 1875, Mr. William Force, of Keyport, N.J., built the steam yacht *Ocean Gem*, for Mr. Rutter, of the Central Railroad. She was 101 feet long, 12 feet beam, and 6 feet draught. Tonnage, 112.83. This was a very successful vessel, and gave entire satisfaction to her owner. She is one of the best of the trunk cabin class.

The following steam yachts were built by firms whose principal business is in freight and passenger vessels:

Messrs. Ward, Stanton & Co., of Newburgh, built three of the most successful steam yachts in the fleet.

The *Vedette*, built for Mr. Phillips Phœnix, in the year 1878. Length, 123 feet; breadth, 18 feet 5 inches; depth, 9 feet 8 inches; tonnage, 191.83.

The *Polynia*, built in 1881. Length, 154 feet 5 inches; breadth, 18 feet 5 inches; depth, 11 feet 6 inches; draught, 9 feet 8 inches, for Mr. James Gordon Bennett, and the *Namouna* built in 1882. Length, 226 feet 10 inches; breadth, 26 feet; depth, 15 feet 2 inches; draught, 14 feet 3 inches, for the same gentleman.

This firm, I believe, built several other yachts, and they furnished the machinery for some of Mr. Lorillard's vessels. They had a high reputation for their machinery, and especially for boilers.

The *Wanda*, built in 1885 by Messrs. Piepgras & Pine, at Williamsburgh. Length, 138 feet; breadth, 18 feet; depth, 11 feet; draught, 10 feet 2 inches. This vessel has great speed, and with the alterations now making, is expected to be the fastest boat of her size in American waters.

The *Ideal*, owned by Theo. A. Have-

"AÏDA," OWNED BY W. P. DOUGLAS, NEW YORK YACHT CLUB.

myer and Hugo Fritsch was built at Williamsburgh, by J. B. Van Deuson, and was launched September 9, 1873. Length, 130 feet; water-line 110 feet; keel, 105 feet; beam, 20 feet 2 inches; depth of hold, 8 feet; draught, 6 feet; schooner rig, 145 tons; engine built Yale Iron Works, New Haven, Conn., two vertical acting cylinder 16 inches by 14 inches; surface condenser boiler 12 feet by 11 feet by 6 feet 7 inches; engine condemned and taken out and new single engine put in; cylinder 20 inches by 22 inches, by Delamater & Co., 1874. Lost, 1884, on the coast of Maine.

The *Ripple*, paddle steamer of the river steamer type, built at Port Jefferson, L.I., for C. A. Chesebrough, of Northport, L.I., is used for cruising with his family in Southern waters during the winter. Length, 118 feet; water-line, 110 feet; beam, 26 feet 3 inches; depth of hold, 6 feet; draught, 4 feet; engine built by Quintard Iron Works, New York, 1880; cylinder inclined direct acting (ferry-boat style), 22 inches by 60 inches, boiler 12 feet by 8 feet, 70 horse power, 85 51-100ths tons, schooner rigged.

Wave (iron), B. F. Lopen, 1864, built in Philadelphia by Reamy & Neafie. Length, 87 feet; 19 feet 6 inches beam; 7 feet depth of hold, 5 feet draught; two cylinder high-pressure 12 inches by 18 inches; propeller, 80 68-100ths tons now owned in Philadelphia.

The America, Henry N. Smith, built by Henry Steers, at Greenpoint, and launched March 1, 1873. Length, 189 feet; water-line, 183 feet 6 inches; keel, 177 feet; beam, 27 feet; depth of hold, 14 feet 6 inches; draught, 12 feet; two cylinders, direct acting, 33 inches by 33 inches; boiler, length, 29 feet 3 inches by 13 feet by 11 inches, low-pressure engine, built by Fletcher Harrison & Company, New York; tonnage, 730, now owned by the Navy Department, and named the *Dispatch*.

I have had numerous inquiries from gentlemen who contemplate owning a steam yacht, as to the expense of running such a

"ORIENTA," OWNED BY J. A. BOSTWICK, AMERICAN YACHT CLUB.

vessel, and I consequently now give them the benefit of my experience on this subject.

The expense of running a steamer depends on a number of circumstances. In the first place, the cost varies with the size of the vessel. A steam launch 40 to 50 feet long requires a pilot, an engineer, and one deck hand. The wages of these ought to be $60 a month for the first two, and scribed in the beginning of this article, there would be required, in addition to the hands mentioned above, a stoker at $40, a cook at $40 (or any higher rate, as the proprietor might choose), a steward at $50, two additional deck hands at $30. Then the pilot and engineer would have to be of a higher class, requiring $80 each, and thus raising the monthly wages to $380. The commissariat would cost at least as

THE OFFICERS' ROOM, BY BOURGAIN.

$30 for the last — making $150 a month. As owners do not usually live on board of such a craft, there would be no expense for provisions, except for the board of the men which might cost $75. Then coal would not cost more than $75, and repairs and sundries, $50 — making an aggregate of $350 per month, or $1,750 for the season of five months. To this must be added the annual refit and clothes for the men, amounting to say $750 more — making a grand total of $2,500.

For vessels of the second class, as de-much more: the coal $200, and repairs and sundries, including the men's uniforms, say $540 — making $1,500 a month, or $7,500 for the season. To this must be added $2,500 for laying up and putting in commission — making a grand total of $10,000.

Vessels of the third class cost very little more than the above, requiring, perhaps, an assistant engineer and one more deck hand, perhaps a mate, thus increasing the total by a couple of thousand dollars.

I now come to vessels of the largest class, which require altogether a different scale

"RADHA," J. M. SEYMOUR, OWNER, AMERICAN YACHT CLUB.

of expense from the others. Such ships as the *Nourmahal* and *Atalanta* carry a complete crew with first and second mates, first and second engineers, double sets of oilers, stokers, deck hands, assistant cooks, numerous stewards, etc. etc. Some of these vessels have been in commission all the year round, and the annual expense of keeping them up must be quite serious.

Such vessels as the *Corsair* and *Stranger* need not cost much more than the vessels of the third class mentioned above. The crew of my boat number seventeen all told, and my consumption of coal is only $350 a month, though I use the boat every day, and the fires never go out except on Sundays.

The expense of running a steam yacht may be kept within very reasonable bounds, or it may be increased indefinitely, according to the way in which it is done. Any one entertaining largely will, of course, run up an important sum for the commissariat, but this I do not consider a legitimate part of the expense of yachting. A yachtsman owning a cottage at Newport might entertain at his house instead of his yacht, and spend the same amount, but it would not then appear as an item in his yachting expenses.

"POLYNIA," W. H. STARBUCK, OWNER.   "STILETTO," BUILT BY HERRESHOFF.

DECK OF THE "NAMOUNA."

"NAMOUNA," JAMES GORDON BENNETT, OWNER.

"NOURMAHAL," OWNED BY WILLIAM ASTOR, NEW YORK.

"WHISPER," OWNED BY R. A. SEACOMB, SEAWANHAKA YACHT CLUB.

I may, therefore, estimate the expense of keeping a steam yacht as follows: First-class launches, $2,500; second-class trunk cabin vessels, $7,500 to $10,000; third-class, flush deck vessels, $10,000 to 12,000; fourth-class such as the *Corsair* and *Stranger* $15,000. The very large vessels, such as the *Namouna*, *Atalanta*, and *Nourmahal*, I have no means of estimating.[1]

Steam yachting is increasing in favor year by year, and there is every indication that it is destined in the near future to be the leading style of yachting in American waters. From very small beginnings, some thirty years since, it has grown in magnitude, until we now have a fleet of vessels varying in length from 40 to 250 feet, and in size from 10 to 1,300 tons. Judging

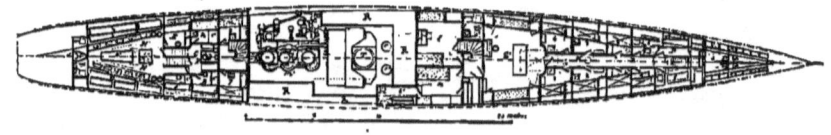

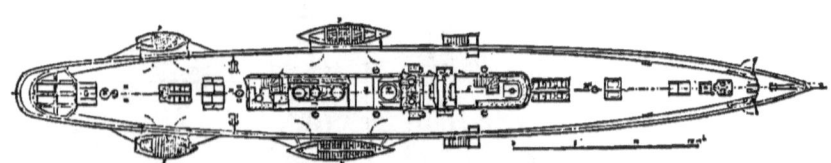

"NOURMAHAL.—WORKING DRAWINGS."

they are much better fed and housed than in sea-going steamers. Yachtsmen, therefore, have the pick of the seamen and most of their employés are Scandinavians. Mates receive from $45 to $100 per month, and engineers the same, while sailing-masters or captains get from $100 to $200 per month.

The steward gets from $60 to $100 a month, and the cabin waiters the same as the sailors. In the galley, the assistant cooks are paid from $40 to $60 per month, but what the chief

---

[1] Mr. Townsend Percy has expressed, on the whole, the most authoritative opinion in this matter, and we beg to reproduce it here from the N. Y. *World* of November 8, 1885. As he has chosen his examples from among the extremely rich yachtsmen, our friends need not be unduly discouraged.—ED.

The sailors and firemen receive on the average about $30 a month, which is more than is paid on steamers in the merchant service, in addition to the fact that

THE BRIDGE.
(Drawn by Bourgain.)

from the progress made in this pastime thus far, we may confidently expect to see a fleet of a hundred steam yachts before another five years have elapsed, and that the enthusiasm for this style of yachting will far outrun that for the old-fashioned sailing craft. In the near future, the probability is that every gentleman residing during the summer within thirty miles of New York, on the shores of the Hudson or East River, will have his steam yacht in which to sail to and from the city. There is no other mode of traveling to compare to it for pleasure and healthfulness. I may here quote the remark of the proprietor of one of the finest of the fleet, when the immense cost of his vessel was alluded to. " My yacht, it is true, has cost a large sum, but it is worth every dollar of it. It has made a new man of me. Before I built it, I was constantly suffering from dyspepsia and other troubles arising from too close attention to business. Now I am a well man."

In concluding these very cursory and imperfect remarks, I can only testify to the unrivaled pleasure and healthfulness of this pastime, and I cordially recommend all who are able to "Join the glad throng that goes *steaming* along," and thus partake of its satisfaction.

E. S. *Jaffray.*

gets is usually a secret, which the cook is too discreet, and the master ashamed, to disclose. The yachts' crews vary from eight to fifty men, according to her size and the service she does, and the ship's pay-roll from $400 a month to $2,500. Add to this the cost of fuel, at about $4 a ton, and repairs, engineer's supplies, such as waste, oil, tools, etc., deck supplies in the way of canvas, cordage, and the like, and the cost of feeding the crew, at an average of at least a dollar a day per head, and the furnishing of the cabin table, and it will be readily understood that the amusement is only within the reach of millionaire. Take, for example, the *Namouna*, owned by James Gordon Bennett, carrying a crew of fifty men; her pay-roll is at least $2,500 a month. Always in commission, the cost of feeding her crew is at least $1,500 a month. Coal and supplies, repairs and the lavishly-supplied table and wine locker of Mr. Bennett, who entertains large parties on her, regardless of cost, and $150,000 a year is a moderate estimate of her expenses.

Mr. Gould does not entertain as lavishly, but his expenses cannot be less than $6,000 per month, and Mr. Astor's *Nourmahal* will only fall short of the expenses of the *Namouna*, and must be $8,000 to $10,000 a month. E. S. Stokes, who only uses the *Fra Diavolo* a few months in the year, spends $20,000 a season on her.

Mr. Edwin D. Morgan's tour of the world in the *Amy* cost a fortune; her five months' trip is estimated to have absorbed at least $50,000. A yacht under a 100 feet long cannot be kept up, even with economy, for less than $1,000 per month, and on most of them at least twice that amount is spent, and this without counting the interest on the money invested, or the annual depreciation in the value of the property.

Pierre Lorillard sold the *Rhada* to Mr. J. M. Seymour for $60,000, and it must cost $3,000 a month to keep her afloat, and the same is probably true of all yachts of her size. As an old-time yachtsman observed, I don't know which will eat a man up the quickest, an extravagant wife or a steam yacht, but think of a rich man with both.

BRITISH YACHTING.

BRITISH YACHTING.

BY C. J. C. McALISTER.

IMMEDIATELY succeeding the prosaic and practical period of the Commonwealth, amongst the numerous sports and pastimes which Charles II. introduced to amuse his subjects, long tired of the restraints of the Puritan rule, was the sailing of pleasure-boats in trials of speed, on the Thames near Lambeth. The Merrie Monarch himself apparently evinced considerable interest in these aquatic contests, which were amongst the most manly and healthful amusements of an essentially effeminate period of English history; but he seems to have utterly failed to succeed in inducing the lords and ladies of his luxurious court to appreciate a pastime attended frequently with so many minor inconveniencies as must have resulted from the climatic conditions under which boat-sailing was and is practised during the short and uncertain season of an English summer. Towards the close of his reign these regattas, inaugurated with the Restoration, which furnish the first record of English yachting, were discontinued altogether. The pastime, as now indulged in, is an institution of comparatively recent date. The first club was formed in 1720, at Queenstown, Ireland, under the title of the Cork Harbour Water Club, which is now known as the Royal Cork Yacht Club. It was not till 1812 that the sister island followed the example through some forty gentlemen establishing a similar association at the Isle of Wight, known as "The Yacht Club," which continued steadily to increase in membership and importance until 1820, when it attracted the attention of William IV., then Duke of Clarence, who ordered that it should henceforth be styled "The Royal Yacht Club," and a few years after his accession to the throne

"GALATEA."

YACHT STATIONS OF THE BRITISH ISLES.

he expressed a wish that it should assume its present title, "The Royal Yacht Squadron," as "a token of his approval of an institution of such national utility."

Doubtless long prior to the establishment of yacht clubs or the introduction of the Dutch term "yacht," there existed ample opportunities for those who had the means and inclination to indulge in amateur aquatic pursuits. The vessels were then called brigs, and sloops, and luggers, and in the older days they frequently sailed for richer prizes than Queen's cups, or hundred guinea purses. When any misunderstanding with England's maritime neighbors led to a declaration of war, there were Spanish or Dutch or French merchantmen, or occasionally a man-of-war of the smaller class, to be sailed after and captured, if the predecessor of the modern yacht possessed sufficient speed, and carried a crew strong enough to board the enemy and bring her back in triumph to an English seaport.

Since 1820, and more particularly during the past thirty years, yachting associa-

"IREX."

tions have made rapid strides in numbers, strength, and popularity. There are now over fifty "Royal or recognized" yacht clubs distributed around the coasts of the United Kingdom, and one at the Channel Islands. To these may be added some fourteen minor associations formed by members owning only the smaller class of craft. The following figures, which have been compiled up to date, will give some idea of the importance of the British pleasure-fleet, as regards numbers, rig, and tonnage :—

Rig.	Numbers.	Tonnage.
Cutters	1,098	15,059
Schooners	285	29,826
Yawls	437	16,566
Steamers	534	53,542
Sloops	57	269
Luggers	55	167
Ketches	5	261
Brigantines	3	1,229
Wherries	5	42
Brig	1	141
Total	2,480	117,102

The "national rig," as the cutter is regarded amongst British yachtsmen, is at present more popular for racing, and also with boats under forty tons, for cruising, than it has ever previously been. This rig appears to have been considered best by racing yachtsmen prior to the advent of the *America*, in 1851; but her famous performance in English waters had the effect of turning the attention of designers to the merits of the schooner rig, which for many years became fashionable both for racing and cruising craft. In fact, it was only a few seasons ago that at many of the principal contests, schooners were found in the majority; but year by year, they have been steadily losing ground. Last season there was only one, the *Miranda*, that entered in the first class matches, open to all rigs, and this year she, also, has hauled down her "fighting flag" and joined the ranks of the ex-racers and cruisers. Judging solely by the evidence afforded through European yacht-racing during recent years, the cutter rig has undoubtedly proved the most weatherly, faster in reaching, and with the assistance of a spinaker, quite able to hold its own with the schooner on a dead run to leeward. Few new schooners are now being built, and although their aggregate tonnage is still considerable, there has been a falling off in their number in the course of the past two seasons, to the extent of fifty-five vessels, while during the same period cutters have very considerably increased. The yawl rig is found convenient for cruising, as by its adoption the

"MIRANDA."

heavy main boom is very much modified, and consequently a yawl of the same size as a cutter can be handled by a smaller crew. Sloops have never been popular in England, and the few that are now sailed are only craft of the smallest size.

The capital sunk in the British pleasure fleet is estimated to amount to over $13,000,000, and independent of amateurs, the number of paid hands required by the yachts is close upon 12,000 men. These are probably the smartest seamen to be found sailing under the British flag; they are in fact a distinct class, and differ materially from the ordinary fishermen and the crews who man the vessels of

"QUEEN MAB."

merchant service. The latter exhibit an embarrassing want of confidence in the long, low, heavily sparred yacht as she lists under a crowd of canvas, till the white water is rushing two or three planks deep over her lee deck, and sending clouds of spray from her weather bow, every time she meets the broken crest of a wave, and in her haste appears to have forgotten to curtsey to it. The merchant sailor who may have been shipped on an odd occasion as an "extra hand" on board a racing craft expresses his objections to yacht racing with a candor and emphasis charactistic of his profession. Accustomed to a large vessel, standing high out of the water, he finds himself on board a craft with bare decks and but little free-board. On the latter he asserts that he is "always too near the water and frequently actually in it." On the other hand the yachtsman who has been all his life used to such conditions of sailing, feels himself perfectly at home. He knows his vessel to be well and strongly built, and ballasted to the nicety of an ounce; that her gear is the best that money can provide, and that his mates are, one and all, to be implicitly depended upon for courage and coolness in any emergency which may unexpetedly arise. He fully appreciates the chances of a spar or some portion of the gear carrying away, and he knows that he may frequently have to spend a considerable number of minutes on a stretch, up to his waist in water, in the lee scuppers, or out on the end of the bowsprit in a seaway. These are merely incidents he is aware he must look forward to, and when they are passed he reflects cheerfully that "they all come in the course of a day's

"ULEDIA."

"CARLOTTA."

work." British yachtsmen of to-day are the result of a long and careful process of selection. They are recruited from amongst the smartest members of the fishing and other seaside portions of the population. They must be steady in nerve, and strong in arm, cool and self-reliant, and amenable to a discipline which is more the out-come of intelligence and mutual confidence than any hard and fast rules or regulations. In the winter season some may, as they term it, "go steamboating" for a trip or two, but they far more frequently, at their native villages, wile away the time with a little fishing or piloting, till the spring comes round and brings with it their season of activity and adventure.

Yacht racing has long been a popular form of sport around the seaports of the United Kingdom, and a considerable amount of money is annually awarded by the various yacht clubs in the form of prizes. Last year, in addition to numerous cups, money prizes to the extent of $62,630 were competed for. The *America's* successes had unquestionably the effect of stimulating interest in the pastime, although in attempting to copy her lines many mistakes were, in the first instance, made by British builders, but the lessons she taught yacht sailors in the art of setting canvas had never been forgotten. A great revolution has during recent years been brought about by the introduction of out-side lead as ballast, which enables the vessels to carry heavier spars with a largely increased sail area.

The most famous British racing yachts of the first class, viz. sixty tons and over, comprise the yawls *Wendur* and *Lorna*, and the cutters *Irex*, *Galatea*, *Marjorie*, *Genesta*, and *Marguerite*. The *Lorna*, now four years old, succeeds in holding her own pretty fairly with the cutters. The *Wendur*, designed by Watson and built two years ago, is considered by many to be the fastest yacht of the fleet. She is built of steel and had seventy-five tons of lead run into the bottom of her keel. As yet she has been but little raced, and although she has more than once proved her speed and weatherly qualties, in the best society, she has been singularly unfortunate in the matter of carrying spars, and losing her leads, through unfavorable shifts of wind. The *Marjorie* is another design of Watson, and was built the same season as the *Wendur*. She belongs to the owner who sent the little ten-ton *Madge* to America, on the deck of an Atlantic liner, some years ago. The *Irex* was a new boat last season, and has proved herself, in every respect, a fast and powerful cutter. The *Galatea* is one of the present season's additions to the racing fleet. She was designed by Richardson, but has so far done little to distinguish herself, as her skipper appears not yet to have been able to find her trim. *Marguerite*, another design of Richardson's, has been sailing remarkably well during the present season, while the merits of *Genesta* are now as well known on one side of the Atlantic as the other. In the second, or forty-ton class, *Tara*, with a breadth of less than one-sixth of her length, built from Webb's design, has had all the best of the racing in her own class this season, besides on several occasions in light weather saving her time from the crack representative of the first-class division. Watson, with *Clara* and *Ulerine*, has designed the two most successful racing boats in the twenty and ten-ton classes, and Paton's little three-ton

"DAWN."

cutter *Currytush*, with even less beam in proportion to length, than *Tara*, has proved herself exceptionally fast in all conditions of weather.[1]

The expenses of sailing a racing craft are in every respect much heavier than a cruising boat of similar tonnage. The cost of building and equipping a ninety-ton cutter of the modern type is $35,000, and in this estimate no allowance whatever is made for cabin fittings, as it merely includes hull, spars, sails, gear, and ballast. The outlay in racing a yacht of this size during the four months' season will amount to quite $10,000. Wages are not a very serious item, considering the class of seamen whose services may be secured. The usual weekly scale is master, $12 to $15; mate, $9; and seamen, $6 50. The crew find their own provisions, but it is usual for the owner to supply clothes, which cost for master, $50; mate, $30; and seamen, $18. These clothes are really a livery, and legally belong to

Name	Rig	Tons	Length	Beam	Depth	Designer	Date
Wendur	Yawl	143	102	17.9	14.3	Watson	1883
Lorna	"	90	83.4	15.9	11.3	Nicholson	1881
Miranda	Schooner	139	92.5	18.8	12.7	Harvey	1876
Irex	Cutter	85	87	15.1	11.3	Richardson	1884
Marguerite	"	63	78	13.6	10	Richardson	1884
Marjorie	"	72	79	14.5	11.6	Watson	1883
Galatea	"	91	90.6	15	13.2	Richardson	1885
Tara	"	40	70.9	11.6	10.1	Webb	1883
Genesta	"	85	85.6	15	11.9	Webb	1884
Clara	"	20	57	9.1	8.5	Watson	1884
Ulerine	"	10	43.2	7.3	7	Watson	1884
Currytush	"	3	31.7	4.9	5	Paton	1884

"CONSTANCE."

"MARJORIE."

the owner, but it is generally the custom to allow the men to keep them at the end of the season. In a racing boat, in addition to the regular wages, the skipper receives ten per cent. upon the amount of the season's winnings, and the men are allowed $5 for every first prize secured.

The Yacht Racing Association[2] was formed in 1875, the object being to provide one code of sailing rules in all matches, and to decide such disputes as may be referred to the council. This association in fact bears the same relation to yachting as the Jockey Club does to horse racing. There is, however, one important difference between the two pastimes. British yachtsmen have not yet learned to demoralize their favorite sport by laying wagers upon the results of races. The yacht-racing season commences towards the close of May, but the nights and mornings are still chilly and the bleak east winds linger with sufficient force to render life comparatively miserable. Although matches and regattas are arranged to take place at all the yachting stations during the season, there is one round in particular, as indicated upon the chart, which is regarded as including all the best sport of the year. For this cruise, which is usually attended by all the fastest and newest boats of the racing fleet, besides a

[2] Yacht Racing Association's method of measuring tonnage.—"The tonnage of every yacht entered to sail in a race shall be ascertained in the manner following: The length shall be taken in a straight line from the fore end to the after end of the load water-line, provided always that if any part of the stem or stern post, or other part of the vessel below the load water-line, project beyond the length taken as mentioned, such projection or projections shall, for the purposes of finding the tonnage, be added to the length taken as stated; and any form cut out of the stem or stern post, with the intention of shortening the load water-line, shall not be allowed for in the measurement of length, if at or immediately below the load line, nor above it within six inches of the water level; the breadth shall be taken from the outside to outside of the planking in the broadest part of the yacht, and no allowance shall be made for wales, doubling planks, or mouldings of any kind; add the length to the breadth and multiply the sum thus obtained by itself and by the breadth; then divide the product by 1730, and the quotient shall be the tonnage in tons and hundredths of a ton."

"IONE."

considerable number of cruisers whose owners take an interest in the contests between the crack vessels; the yachts assemble on the estuary of the Thames, off Gravesend, Erith and Southend. There is good and safe holding ground, but the river in its lower reaches neither furnishes picturesque scenery nor, owing to numerous sand banks, narrow channel and crowded traffic, a satisfactory course for fairly testing the respective merits of the competing vessels. After some half-dozen matches, sailed under the auspices of the numerous yacht clubs, with stations in this neighborhood, June opens with a channel match from Southend, at the northeast entrance of the river, over a course of forty-five miles, to Harwich. This portion of the Essex coast, which is passed, is low, bare and uninteresting, but Harwich itself well repays a visit. It is an old-world port with no trade, and bears all the appearance of having been asleep for the past hundred years. But Harwich is a place with a history. The pilot who guided the fleet of little ships in which Julius Cæsar crossed from the coast of Gaul to Britain fifty-five years before Christ is said to have been a Harwich man, and here during all the time of the Roman occupation a strong hold was maintained to repel the attacks of the Danes and Saxons; and it was from Harwich that Edward III. embarked in 1338 on board a fleet of 500 sail manned with archers and slingers on his first expedition against France; and during later years the English had many a stout encounter with the Dutch and French fleets within sight of this quaint old Essex port. The place itself to-day bears far more the aspect of a Dutch than an English town. There is the level coast line, and dykes, and wind-mills, and red-tiled houses, that one is accustomed to look for only in Holland. Here, within the estuary of the Stour and Orwell rivers, the Royal Harwich Yacht Club provide an excellent day's sport over a course where the breeze blows steadily above the low shores, and the tides are not sufficiently strong to interfere with fair sailing.

Southwards the fleet returns past the mouth of the Thames on a channel match to the headquarters of the Royal Cinque Ports Yacht Club at Dover. Most people who have visited Europe are familiar with all that is of interest in connection with this comparatively modern Kentish port. The course is right out in the channel, and is usually sailed over in a strong breeze with a lumpy sea, which the cross-tides create. The white cliffs and the old castle are perhaps best seen from the bay, but even they render anxious moments for the smartest crews. There is no really good racing to be obtained here, as from beginning to end of a match it is generally only a matter of working the tides. Many of the cruising boats, and frequently a few of the racers, instead of venturing up the Mersey run over to the Isle of Man, and, weather permitting, come to anchor for a day or two in Douglas Bay. The Manx capital during the summer season is a bright and cheerful little town, and there are interesting trips to be undertaken in different directions over the island during the short stay. Northwards the racing fleet steer to Mor-

"DIANE."

will hardly induce the yachtsman to wish to prolong his stay beyond the two days occupied with the regatta. From Dover the fleet have before them the longest trip of the cruise. Away down the English channel to the westward, round Land's End and up St. George's channel to Liverpool. The Mersey is only known to most yachtsmen to be avoided. In the course arranged by the Royal Mersey Yacht Club, the start takes place just abreast of the Princess' Landing Stage, associated in the minds of most Americans only with their arrival or departure from Europe. The strong tide which rushes in and out of the Sloyne, and the crowd of vessels anchored in the stream, cambe Bay, to attend the regatta of the Royal Barrow Yacht Club. The town is merely a manufacturing place of very recent growth, but it is a convenient point from which to reach Furness Abbey and the charming scenery of the English Lake district. From Barrow a course is shaped round the Mull of Galloway and up the estuary of the Clyde. It is high midsummer by the time the fleet reach Scottish waters, and during the whole course of the short night in these high latitudes, the daylight never quite fades from the sky.

During a somewhat varied experience, I have spent nights on the Bosphorus, and sailed under the Eastern moonlight up the

Golden Horn. Many a night has been passed under the shade of the palm trees, in the coral lagoons of the South Sea Islands, watching the flashing torches of the native fishermen. From the entrance to the Golden Gate I have seen the sun waters flash pink and yellow under the reflection of the fading sunset, or the first rays of the sunrise, and the peaks of the Arran and Bute Mountains loom—tinted with dreamy purple and blue—against the bright hues of the western clouds. No-

"BUTTERCUP."

sink into the broad Pacific, and I have experienced the pleasure of steering an open boat by moonlight up the head-waters of comparatively unknown rivers in New Zealand; but nowhere have I so thoroughly enjoyed the witching hours as whilst yachting during the soft midnight light, on the estuary of the Clyde. The land-locked where around the British coasts is yachting regarded with keener enthusiasm than by the dwellers on the shores of the Clyde. The numerous yacht clubs provide a lengthened programme of events, which usually occupy the best portion of two weeks; but the shores of the Scottish river are much too high and picturesque to permit the breezes to blow as steadily as every one could wish, in the interest of fair sailing, and many a good topmast has come to grief before the sudden and unexpected gusts that come sweeping down the glens.

From the Clyde the fleet cross to Bangor, Belfast Lough, which is the first Irish part touched on the cruise. Here two days have been arranged by the Royal Ulster Yacht Club over one of the best courses in the United Kingdom. Bangor is a modern and badly laid out seaside resort; but on the opposite side of the lough the gray old castle of Carickfergus bears testimony to many a hard fight, in days gone by, between the Scotch and Irish chieftains. On departing, a southerly course is followed by the fleet to Kingstown, Dublin Bay, where the St. George's, Royal Irish, and Royal Alfred Yacht Clubs provide a sufficient number of matches to involve a week's stay. There are only two conditions of weather which appear to usually prevail in Dublin Bay during regatta time. One is a flat calm, and the other a strong southeast breeze, which sends a great rolling sea into the bay, which makes lively times for the smaller craft in sailing over the exposed course. The next port made for is Mumbles, on the Bristol Channel, where several days' racing is given by the Bristol Channel Yacht Club; and then the fleet round Land's End again to Falmouth, where they race for the prizes given by the Royal Cornwall Yacht Club. These matches are immediately followed by the regatta of the Royal Western Club, at Plymouth. There is no prettier course in English waters than the one sailed out through old Plymouth Sound. The area within the breakwater is close upon twenty-five miles, bounded on the west by the richly-wooded heights of Mount Edgecumbe, and on the east by Mount Batten and the Wembury cliffs. The Hoe, an eminence near the town, is where the stout

"TARA."

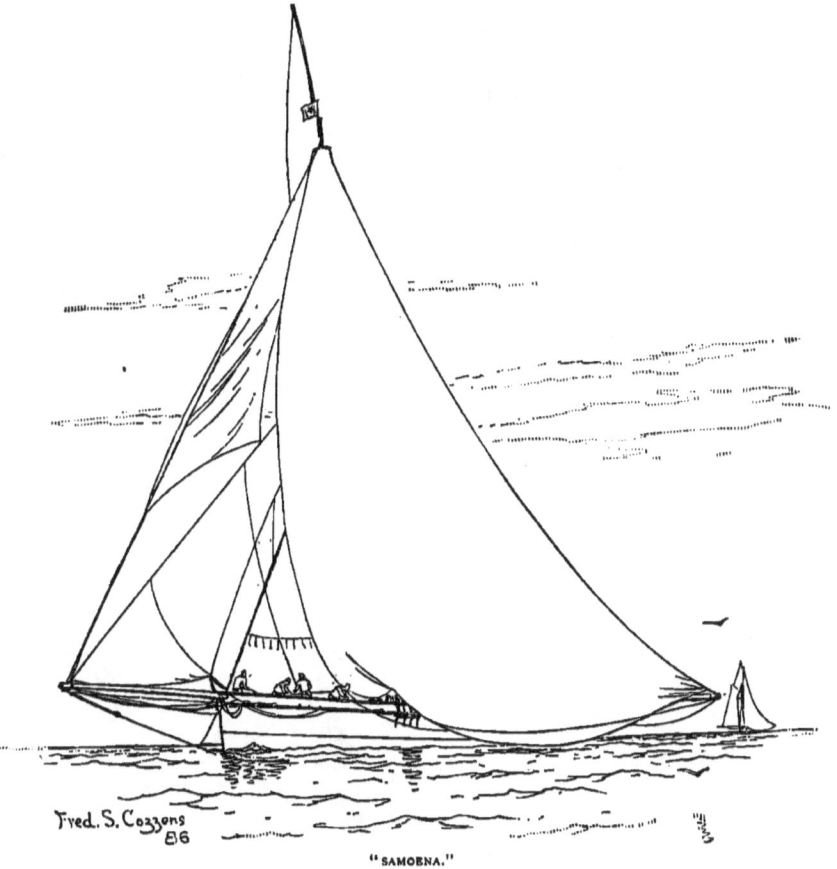

"SAMOENA."

old English admiral, Sir Francis Drake, was engaged in playing a game of bowls when he received intelligence of the approach of the Spanish armada. Some of the ships entered the sound, and their admiral, the Duke of Medina Sidonia, is said to have been so much pleased with the situation of Mount Edgecumbe that he determined to make it his residence when the forces under his command had conquered England. Sir Francis persisted in quietly finishing his game of bowls; putting to sea with the little English fleet, to play the sterner game, which—with the help of a storm—ended in the destruction of the Spanish ships. On regatta days the Hoe is crowded with a concourse of people, who look down on the same land-locked water that harbored the little hundred-and-eighty-ton *Mayflower* before she shook out her sails to the breeze, just about 300 years ago, to start upon her memorable voyage to the shores of New England.

The next event of importance for the racing fleet is the regatta of the Royal Yacht Squadron at Cowes. This is by many regarded as the great aquatic carnival of the year. Nevertheless, the racing is usually of the tamest possible description, as for several of the events—notably, the race for the Queen's Cup—members only are permitted to enter. This club is perhaps the least representative of all the yachting associations in England, and if the interests of the pastime were left solely in its hands, yachting would have sunk to a low ebb indeed. The members form a curious combination of "swells" and "snobs." The former, with the Prince of Wales at their head as commodore, are generally smart yachtsmen, but the latter, unfortunately, preponderate, which probably accounts for the curious manner in which the affairs of the club are managed. The Royal Squadron appears to be regarded rather in the light of a joke by the members of the exclusively aristocratic, but more enterprising associations. Sir Richard Sutton belongs to this club, but the *Genesta* is the only boat out of the entire fleet that would have the faintest chance in a race open to all comers. Cowes is gay enough during regatta week. To the right of the bay is the club-house, a queer little gray, ivy-clad building, that looks as if it had been built as a model for some important establishment, and proved a failure. Across the Madina River, on the summit of some high, wooded land, is the royal residence of Osborne, and out across the Solent—which is blue, occasionally, when the weather is clear—are the low, level shores of the Hampshire mainland, and beyond Southampton Water. The Solent is frequently well filled with merchant steamers, men-of-war, and sailing vessels, making their way up or down the channel; but as there is plenty of room, the races are seldom interfered with by their presence. A description of the Solent —written eighty odd years ago—describes it as being occupied by a very different

"EGERIA."

class of craft from those found on its waters to-day. "Coasting schooners, fishing-smacks, brave Indiamen, and now and then a fighting-ship—king's, or foreigner, and here and there a sullen-looking lugger, upon which the smart active men of the 'Rose' cutter seem most diligently to wait upon, make up the show of shipping."

After the termination of the Royal Squadron regatta the racing boats are fully occupied at the various ports in the immediate vicinity of the Isle of Wight until the close of the season, which takes place about a month later, when the weather, which is somewhat uncertain during the summer, assumes as the autumn days draw in, a more defined character for the worse.

The cruise I have briefly indicated would probably prove an attractive outing for most American yachtsmen. They would be certain of receiving a most cordial reception and would find the great majority of the yacht clubs placed at their service during the cruise. An opportunity would be afforded them of seeing the United Kingdom and its inhabitants from a totally different standpoint from that occupied by the ordinary tourist. There

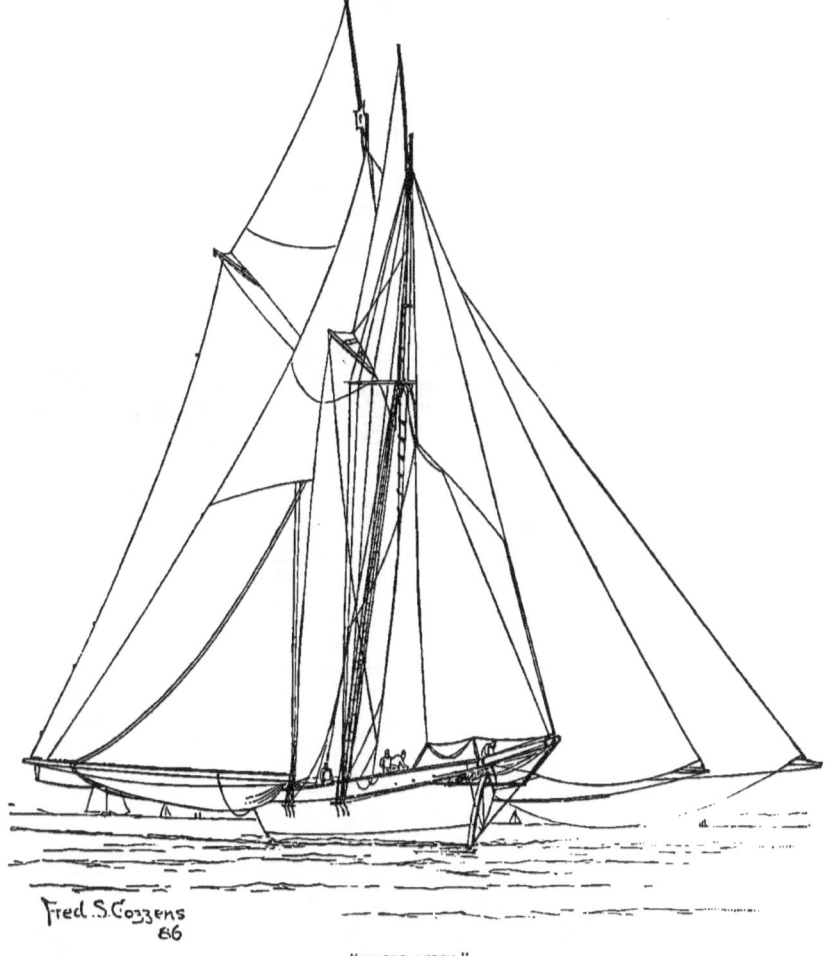

"WATER WITCH."

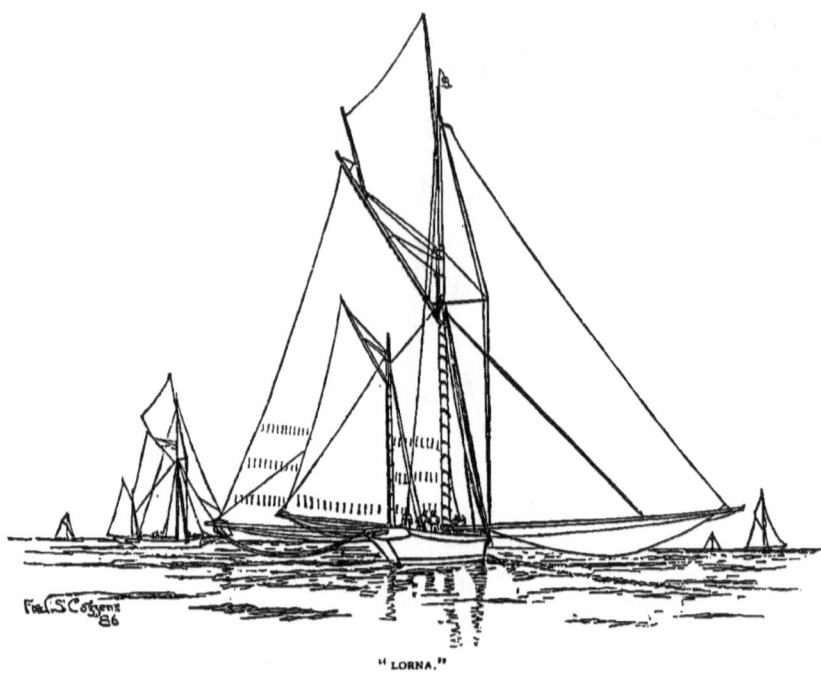

"LORNA."

exists in all parts of the world a certain description of Freemasonry amongst yachtsmen, and nowhere is it more apparent than in British waters. The voyage across the Atlantic is not a very serious undertaking for the larger class of American yachts, and the smaller boats can be placed on the deck of a steamer and sent over without any very serious outlay. To all keel-built boats belonging to yacht clubs, the races

"GERTRUDE."

are open, and the American owner might, if his craft prove fast enough, manage to return with a whole locker full of silver cups, and hundred guinea purses. But there is no necessity to race. Abundant interest and amusement may be obtained on board a cruiser of very moderate tonnage, and if it is not convenient to send a boat across, plenty of suitable yachts may be hired in England at a very reasonable rate per month. Sir Richard Sutton's trip calls for a return of the visit, and, although he was unsuccessful in his attempt to win the America's Cup, he will, doubtless, with many of his countrymen, be consoled with the reflection, "'Tis better to have sailed and lost than never to have sailed at all."

www.ingramcontent.com/pod-product-compliance
Lightning Source LLC
Chambersburg PA
CBHW030315170426
43202CB00009B/1006